DEVICES THAT ALTER PERCEPTION
2010

DEVICES THAT ALTER PERCEPTION 2010

First Edition

Devices that Alter Perception 2010

Editor: Carson Reynolds
Design: TheGreenEyl
Printing History: June 2011: First Edition
ISBN-13: 978-1463664244
ISBN-10: 1463664249
Cover Photo: Alvaro Cassinelli, EARLIDS, 2008

CONTENTS

DEVICES THAT ALTER PERCEPTION PREFACE

WORKSHOP CONCEPT With the desiccated carcasses of design concepts littering the world like cicada shells, it takes a bit of nerve to structure a workshop around a new one. However, we have found over three years with accompanying workshops that the concept of devices that alter perception has stuck and continued to motivate us. What follows are excerpts from this year's call for papers. We hope this provides some context for the abstracts and imagery that comprise this booklet.

CALL FOR PAPERS ABSTRACT Sensors, actuators, implants, wearable computers, and neural interfaces can do more than simply observe our bodies: these devices can also alter and manipulate our perceptions. This workshop will promote the design and critique of systems whose explicit purpose is to alter human percepts. Participants will be asked to present abstracts, images, videos and demonstrations that focus on devices that shape perceptual phenomena. The goals of the workshop are to:

(1) document an emerging field of device design;

(2) facilitate the development of these devices by sharing designs;

(3) better understand the process of perception and how it informs the design of devices; and

(4) debate the aesthetics, perceptual change, social and ethical issues as well as functional transformation the presented works envision for the future.

BACKGROUND Systems that perform sensory substitution[1] as well as techniques like galvanic vestibular stimulation[2] (which interact with human balance and gait) demonstrate actual devices that act upon human percepts. Projects such as Body Mnemonics[3], which makes use of proprioception, the FeelSpace belt[4] which offers a variety of magnetic perception, Haptic Radar[5] which augments perception of space, and Low-fi Skin Vision[6] which illustrates sensory substitution all provide stronger evidence of a growing genre of on-human perceptual devices.

Philosophically, we are keenly interested in accounts of perception and its relationship to tools and devices. Noë's account detailed in Action in Perception[7], we find deeply influential. His development of J.J. Gibson's view of perceptual systems[8] allows us to think theoretically about how the percepts make use of devices ready-at-hand. Clark anticipates this view in Natural Born Cyborgs and with theories concerning human-machine symbiosis[9].

We are further inspired by Stelarc's performances illustrating the relentless hybridization of human and technology[10].

The augmented reality research community is keenly aware of the importance of perceptual biases such as those discussed by Drascic and Milgram[11]. Instead of viewing these as a human factors problem, we will advocate using such biases and perceptual illusions to create new devices that more tightly integrate with our perceptual processes.

MOTIVATION We wish to develop and bring together work related to human perception such as sensor systems, physical computing, and interaction design projects. The areas of augmented cognition, augmented reality, subliminal user interfaces, brain-computer interfaces, prosthetic design, affective computing and haptics are all overtly relevant. Examples of suitable position paper topics include:

→ Devices which initiate reflexive responses in users
→ Phenomena such as perceptual illusions which can be exploited by systems and devices
→ Media art that makes unconventional use of the viewer's percepts
→ Systems which seek to alter user behaviour subtly
   (such as alerting the user without diverting attention)
→ Prosthetics that transform perception by making use of techniques such as
   sensory-substitution
→ Sensor systems that regulate or reshape emotions
→ Psychological and physiological studies that relate to the process of perception
→ Device designs capitalizing on neuroscience established tools such as EEG,
   but also emerging techniques such as diffuse optical tomography and
   transcranial magnetic stimulation
→ Displays that allow atypical perceptual experiences
   (such as temporal distortions, out-of-body experiences, etc.)
→ Worn devices that simulate synesthesia combining haptic visual and optical sensations
→ Devices that map imperceptible phenomena onto the percepts
→ Interfaces which target awareness by inducing drowsiness or alertness
→ Haptic devices that allow re-experience of another persons' percepts

Carson Reynolds

These examples lead to important questions for designers: Can users control devices that manipulate the percepts? What would be necessary for users to trust these devices? Such ethical design questions deserve consideration and debate since delusory experiences and unreliable perceptions may be unplanned byproducts of these devices.

KEYNOTE This year's dap workshop featured a keynote by James Auger, who is a tutor in the Royal College of Art's Design Interactions department and partner in the Speculative Design practice Auger-Loizeau (www.auger-loizeau.com). He spoke on the topic of Designing Devices that Alter Perception.

Carson Reynolds

References [1] Kaczmarek, K. A., Webster, J. G., Bach-y Rita, P. and Tompkins, W. J. 1991. Electrotactile and vibrotactile displays for sensory substitution systems. IEEE Transactions on Biomedical Engineering, vol. 38, no. 1, pp. 1–16, January 1991. [2] Maeda, T., Ando, H., Amemiya, T., Nagaya, N, Sugimoto, M., and Inami, M. 2005. Shaking the world: galvanic vestibular stimulation as a novel sensation interface. In: ACM SIGGRAPH 2005 Emerging Technologies (Los Angeles, California, July 31 - August 04, 2005). D. Cox, Ed. SIGGRAPH '05. ACM, New York, NY. [3] Angeslevä, J, O'Modhrain, S., Oakley, I., and Hughes, S. 2003. Body mnemonics. In: Mobile HCI Conference 2003, volume 16. [4] Nagel, S. K., Carl, C., Kringe, T., Märtin, R., and König, P. 2005. Beyond sensory substitution—learning the sixth sense. Journal of Neural Engineering, 2(4):R13+. [5] Cassinelli, A., Reynolds, C., and Ishikawa, M. 2006. Haptic radar. In: SIGGRAPH '06: ACM SIGGRAPH 2006 Sketches, New York, NY, USA. ACM. [6] Bird, J., Marshall, P., and Rogers, Y. 2009. Low-fi skin vision: a case study in rapid prototyping a sensorysubstitution system. In BCS HCI '09: Proceedings of the 2009 British Computer Society Conference on Human-Computer Interaction, pages 55–64, Swinton, UK, UK. British Computer Society. [7] Noë, A. 2006. Action in Perception (Representation and Mind Series). The MIT Press. [8] Gibson, J. J. 1983. Senses Considered As Perceptual Systems. Waveland Press. [9] Clark, A. 2003. Natural-Born Cyborgs: Minds, Technologies, and the Future of Human Intelligence.Oxford University Press, USA, first edition. [10] Stelarc, 1991. Prosthetics, robotics and remote existence: Postevolutionary strategies, Leonardo, vol. 24, no. 5, pp. 591–595. [11] Drascic, D. and Milgram, P. 1996. Perceptual Issues in Augmented Reality. Proc. SPIE Vol. 2653: Stereoscopic Displays and Virtual Reality Systems III, San Jose, California, Feb. 1996. 123–134.

DESIGNING DEVICES THAT ALTER PERCEPTION

Around 15'000 years ago our early human ancestors befriended the Wolf. Ortega y Gasset called this alliance: 'The only effective progress imaginable in the chase'[1].

If we understand perception to be an animal knowing its environmental niche then in placing his nose at mans disposal the Wolf became the original device to alter our perception: extending the human sense of smell far beyond natural capability.

Until quite recently perception altering devices were based in similar survival strategies, enhancing natural capabilities to offer advantage in the specific environment in which the device was applied. As technology has progressed, things have become more complicated, survival is no longer a major worry for most of us and altering our perception of environments has entered a far more complex realm.

We now exist in many altered forms of reality, live artificial second lives; trick, cheat and titilate our perception; electronically augment knowledge of selves and environments. With this plethora of possibility should come grand questions, even simply asking why is helpful. In naming this research strand Device that Alter Perception, rather than enhance or augment, the motivation behind the alteration remains vague. Thus the devices created and described here can equally be used to explore and critique our ideas and plans for future devices as much as propose them.

A common approach in techno-centric domains is to focus predominantly on application and function for enhancement purposes, ignoring the contextual factors and implications. If we remove the connection to everyday life when researching a technology, the experiences or the affects facilitated by the device risk being purely a showcase; a kind of fairground ride where the aesthetic experience, however striking, is simply a temporary alteration of reality. This can create intrigue, thrill and fascination but the effect is rarely enduring, existing like a one-line gag.

Alternatively, potential dark or disruptive consequences of application pass unnoticed or unchallenged due to the removal of real-life sensibilities. Taking inspiration from James J. Gibson and his statement: 'It is not

James Auger

Auger-Loizeau, Social Tele-Presence, 2001

true that 'the laboratory can never be life like.' The laboratory must be life like!'[2] The question that must be posed is: where, why, how and when perception altering devices are applied; the contextual issues that can turn a technology into a product and in turn modify the real-life human experience: Where can the alteration of perception lead to more immersive or rewarding mediated experiences; where should the limits lie in technological mediations, when does technology become too invasive?

James Auger

References [1] Gasset, Ortega. Y: Meditations on Hunting, Wilderness Adventures Press Inc. 1995. [2] Gibson, James. J: The Ecological Approach to Visual Perception. Psychology Press, New York. 1986.

OPTICAL HANDLERS – EEYEE

SHORT ABSTRACT The project Optical Handlers – eeyee investigates how one's embodied visual experience can be subverted by the manipulation of visual technology, and explores the realm of perception transformation under the circumstance of visual alternation. With particular deviation of visual experience, the project challenges one's inherent manipulation of senses and body, and encourages relearning and re-appropriating the body.

With a long history, optical devices such as camera have augmented human vision in different aspects to allow human to see with more detail, farther, richer (e.g. Infrared), and so forth. The common quest of these sorts of research is to reveal/discover the unseen/unperceived from the limited naked eye. The core investigation of Optical Handlers – eeyee falls into the portability of vision down to the sensational relocation of vision. Wearable optical device is developed in this project; nevertheless, it does not only make altered vision wearable, but sharable and playable.

OVERVIEW Optical Handlers – eeyee is an optical device that dissects one's embodied visual experience with a simple tool-set – LCDs and cameras. eeyee literally splits one's vision into two and relocates them onto his/her hands separately. The extended vision, along with the mobility of the hands, makes it possible for the user to observe the world from a dimension off the skull and from a perspective through the limbs.

TECHNOLOGY The device comprises two parts: 1) the goggles and 2) the gloves. Within the goggles, two pairs of LCDs are arranged side-by-side in front of the right and left eye separately. Each eye sees TWO LCD images magnified by lenses, which are perfectly aligned in a binocular distance. Eventually, user naturally overlaps four images, which enable dual-stereoscopic vision. All the LCDs' videos are sourced from two sets of stereoscopic cameras on the right and left gloves separately and accordingly, which enables the user to see through his/her hands. An identical duplication of the LCD-set that has the same video sources is equipped in the face of the goggles, facing outward to the audiences. It allows bystanders to see through the user's modified perspective. The cameras also offer another four video signals that connect to a video splitter, which is used for broadcasting and recording the vision in real-time.

Eric Siu, Carson Reynolds

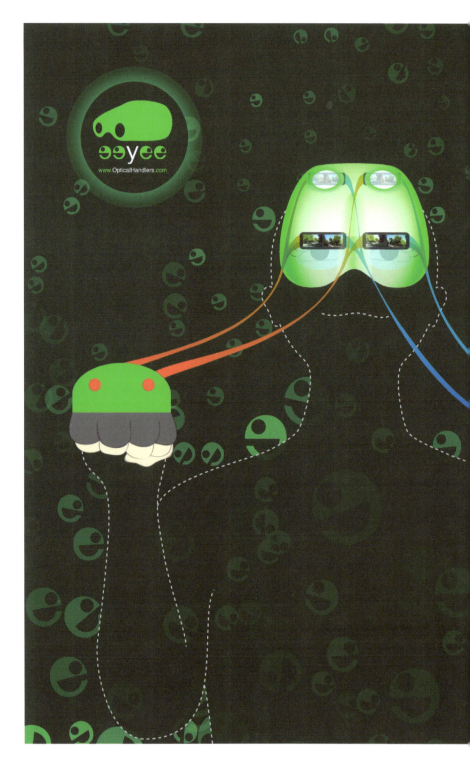

Stereoscopic Vision

Vision Mobilized

Vision Split Stereoscopically

Vision Shared

MOBILE AND DUAL VISION When user wears this device, the experience is completely alienated. Employing the capability of arms and hands, the device heightens the mobility of the vision. The directions your hands able to reach, are now the views your eyes can see. Therefore, even though the user stay put, he/she can see viewing angles that expanded from rolling eyeballs and turning head. Thus, scenes behind the skull, view under the jaw, and most unreachable viewpoints are now reachable by the hands. Human has a pair of eyes, but we normally just perceive one image at a time. Nevertheless, each eye is one independent reception organs. This project makes use of this dual reception organ to create a dual vision. With two mobile visions in hand, it changes the way the user sees and perceives the space. It demands intensive brain process and sometime confuses the user, but once one gets use to it, the dual vision become useful by offering you multi-dimensional viewpoints. The mobile and dual vision requires the user to relearn the body and to invent new ways to cope with their sur-roundings as primitive as making a step forward.

MEDIATED VISION Now, the user perceives through a set of cameras. Unavoidably, the camera mediates the vision by a 4:3 frame space, field of view, focus and exposure. Thus, vision is limited to the capability of the cameras, and the user no longer perceives a wide-open field of view from the eye, but a set of 4:3 frame spaces cropped by the cameras. Hence, the user loses most of the spatial information that supposes to assist him/her to deal with the environment and to protect him/her from danger. Normally, in such a situation and under such a limitation users will interact with the surroundings with caution. The anxiety raised from the limitation causes hesitation in decision-making, and requires tries and errors. To obtain spa-tial information, user has to constantly point at the directions that draw his/her attention. It turns out generate a very interesting behavioral phe-nomenon unique to this project – selective vision.

SELECTIVE VISION The selective vision is highly driven by human senses and environment – hearing, touching and the proximity between user and objects. Since the field of view is limited, user has to integrate other senses with the visual sense to understand the environment. Sound and things that the user bumps into draw his/her attention. The user reacts to whatever stimulates him/her by using the eye(s)/hand(s) to look for the attractions. In other words, user selects things to see by gesturing, which is different to the normal reactions like eyeballs rolling or head turning. Eventually,

Eric Siu, Carson Reynolds

8

it heightens the awareness of the body and complicates the reflexive responses of the user.

ALIENATED STEREOSCOPIC VISION AND HYBRIDIZATION OF SENSES

Instead of splitting the vision into just a left and a right, eeyee treats each split stereoscopically. Thus, the user experiences a double mobile real-time stereoscopic vision. The stereoscopicity of eeyee provides a familiar sense of depth to the user however, relocated. Normally, you use to determine the head-to-objects distance by the binocular vision, which allows you naturally touch objects. However, with wearing eeyee, simple acts such as touching things right in front of you become highly challenging, as you are actually perceiving the point of view shots from your hands. Seeing and hand gesture are therefore woven together. Two circumstances arise in this situation. First, since the sense of depth is extended to the hands, user has to adjust his/her understanding of distance between self and external objects. Secondly, the sense of vision and of touch is laterally combined; it hybridizes the manipulation of both senses. Again, the user has to relearn or invent new ways to use this pair of eyes and hands.

SOCIAL INTERACTION AND VISION SHARING

The device also acts as a performance tool for social interaction. eeyee, is the performer who wears the device in the performance. There are two peeking holes in the face of the goggles that allow bystanders to see what eeyee is looking at. eeyee walks on the street like an alien, awkwardly moves around and learns about the space, and unexpectedly confront people, which then draws tremendous amount of attention and curiosity. It bursts the social bubble of strangers with its friendly and funny alien look. Once someone is encouraged to come inch close to peek in eeyee's eyes, laughter follows. Optical Handlers – eeyee not only makes possible to transform one's visual experience, but also makes the altered vision sharable by a simple action – peeking. Bystanders, while interacting, can move eeyee's hand to enjoy the mobility of vision.

PRESENTATION AND DEMONSTRATION STRATEGY

As a performance, eeyee has gone through various cities including New York (USA), Shanghai (CH), Wroclow (PL), Belfast (IR) and Yokohama (JP). All performances are documented with video and photography, which will be included in the presentation. The device can be simply setup and worn therefore, the audience members can be invited to experience the project. Additionally, since the cameras signals can be broadcasted (to the projection), audiences can observe what the user is seeing in real-time.

Eric Siu, Carson Reynolds

OBJECTS FOR OUR SICK PLANET

Kian Peng Ong

ABSTRACT Objects for Our Sick Planet is a series of artistic works created in response to the various environmental problems that we face. In this paper, I bring forth the argument that humans are disconnected to these real world problems through various reasons such as information pollution and desensitization through mass media. Objects for Our Sick Planet is an attempt to investigate and explore alternate ways of perceiving these problems in ways that are more immediate and relational to our Self.

INTRODUCTION As the world population and its industrial needs continues to grow at an ever increasing rate, environmental problems become a pressing issue at hand. However, a vast majority of us continue to live their everyday lives without stopping to think about the bigger problems such as climate change and how it can affect them or their future generation. One reason for this could be information pollution[1], complex and often abstract data makes it difficult for the typical layman to make sense of such information without first doing some research. Another reason could be frequent documentation of related events by the mass media, such real world events are often far away from viewers watching it in the comfort of their homes. The repetition of such images can cause desensitization[2] and the passive role that we as the audience takes creates a further distance between the event and ourselves. This disconnectedness is what Objects for Our Sick Planet seek to reconcile by attempting to present these problems in ways that are experiential and perceivable without prior technical knowledge.

OBJECT #01 AIR MONSTERS Air Monsters is a toy instrument that investigates the air content in the environment that we live in with a playful approach. This is done through the process of capture, analysis, and visualization. Harmful elements in the air are visualised as animated monsters that have a life of their own, whereby mundane data is translated into a metaphorical representations.

BACKGROUND This work is a direct response to the representation of pollution measurement index used in many relevant agencies and broadcasted to the public[3]. Constant exposure to numbers that fluctuates without having any direct impact in our everyday lives could numb our ability to assess the real nature of what these numbers could actually mean.

NUMBERS AS MONSTERS Travelling through different areas, the user can randomly decide to take a sample of the air surrounding him or her. The Bag like object acts like an accordion that can be deflated and inflated, following which new air sample can enter into this apparatus with an air sensor within. Upon analyzing the air content for possible contaminants, these values are then translated into the form of monsters. The idea of this is to make the otherwise mundane data more meaningful with a point of reference in thinking about the air quality in a particular area.

OBJECT #02 FLOOD HELMET The Flood Helmet is a wearable device that visualises possible future flood scenarios based on the user's physical location. The flood level inside the helmet is determined by the GPS location and elevation height of the land that the user is standing on. Based on the IPCC[4] sea level rise predictions, the Helmet is flooded accordingly. Taking the public's nonchalant response to the possible sea level rise in the next 50 to 100 years[4] and the inability of experiencing any real impacts that sea level rise can possibly bring as a point of departure, Flood Helmet attempts to manifest the possible future in an experiential and immediate way.

PLACING SELF IN CONTEXT Flood Helmet allows a bodily experience that places our Self[5] in relation to the projected future and the location. Taking Tuan's[5] notion of space and place in a central concept in this work, Flood Helmet aims to help audiences rethink about the environment that they are so used to by over-layering the real environment with the Helmet interface. When one puts the helmet on, it encloses them in a small transparent visor that separates them from the direct environment. Instead, the user can only look at surrounding through this helmet. The initial feel of wearing this helmet is one of ease, though not necessary comfort.

As users travel to different locations, they realise that at certain locations water start gushing in from the rear of the helmet. Transforming the view of the landscape with this water layer, it is a simple visualization of data that puts the user in the actual site and experiencing the possible future at once. As the water level increases, the comfort level goes down as well as the weight of the water increases the pressure sitting on the top of the user's head. This is interesting because it adds an extra sensory perception to the user that helps to shape the experience of the work.

OBJECT #03 BRIGHT NOISE Bright Noise is a mobile field survey instrument that scans the light environment in the dark. This work deals

Kian Peng Ong

Top: Flood Helmet

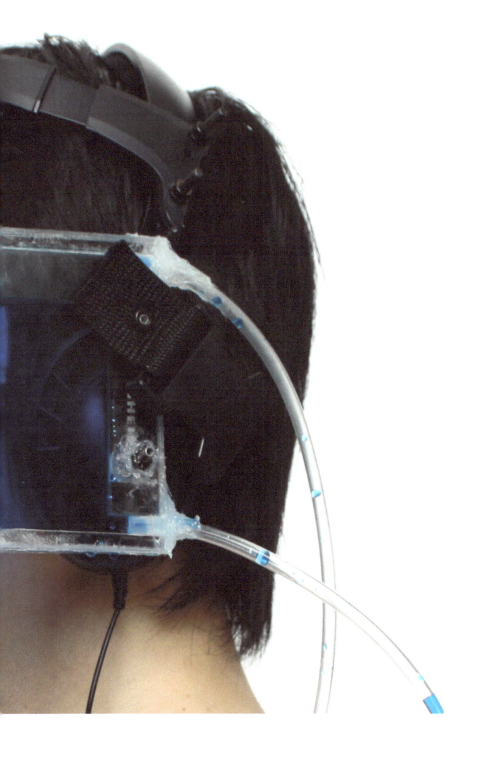

13

Top: Bright Noise, Bottom: Air Monsters

with light pollution and the current problems in using visual means to represent light pollution. The scanned image is converted in realtime into a soundscape of varying intensity and frequency that represents the state of the light environment in the scanned area.

PARADOXICAL REPRESENTATION This project deals with the representation of light pollution in many art works or scientific studies. In many of such works[8], light pollution is represented through a long exposed image of a long shot at the skyline or city. Very often, it looks nice and aesthetically pleasing rather than a problem that needs to be highlighted. Also, historically light has been a symbol of the good and holy[9][10][11] so there exists a paradox in using light based art to represent light pollution.

Instead, I propose sound as an alternate way of perceiving lights. Bright Noise is a mobile field survey tool that scans the light scape of an environment, the result of this is done with a wireless webcam embedded in the tool which does a slit scan. The scanned image is then pixelated into color blocks in which each block carries a number value of the light source. A program is then activated to scan through each of these pixel blocks and each value would then be translated into a sound. Clicks represent dimmer lights and noise represents bright lights. Silence represents darkness. The entire soundscape would represent the light pollution level of an area.

References [1] Fabrikant, S, I. and Buttenfield, B, P. (2001), Formalizing Semantic Spaces for Information Access [online], p.263-280, Available from: http://www.jstor.org/stable/3651259 [Accessed: (21.04.2008)]. [2] Sontag, S.(2002), On Photography, Penguin: London. http://app2.nea.gov.sg/psi.aspx [3] Singapore National Environment Agency PSI Readings http://app2.nea.gov.sg/psi.aspx Retrieved September 2010 [4] IPCC Fourth Assessment Report : Climate Change 2007 http://www.ipcc.ch/publications_and_data/ar4/wg2/en/content s.html Retrieved December 2008. [5] Benson, C., 2003, The Cultural Psychology of Self : Place, Morality and Art in Human Worlds, 5th edition, Routledge: London. [6] Tuan,Y.F., 2007, Space and Place : The Perspective of Experience, 5th edition, University of Minnesota Press: Minneapolis. [7] Neudecker. M. 2008, Think of One Thing http://www.formatnetwork.com/mariele/neudecker.htm Retrieved December 2008. [8] Blake. G. Cloud Projections http://www.blakegordon.com/stories/cloudprojections.html Retrieved Febuary 2009. [9] Weightman, B.A., 1996, Sacred Landscapes and the Phenomenon of Light, American Geographical Society, Geographical Review, America, Vol. 86, No. 1, pp. 59-71. [10] International Bible Society (1984), Book of Genesis [online], genesis 1:1-2:3, Available from: http://www.biblegateway.com/passage/?search=genesis%201: 1-2:3&version=31 [Accessed: 22.4.2009]. [11] Pappas, N., 2003, Routledge Philosophy Guidebook to Plato and the Republic, 2nd Edition, Routledge: London.

Exhibition: All works can be viewed online at http://ctrlsave.com/ofose/

AAA-AUTOMATIC ANCHORING ARMOUR INSTANT THERAPY FOR NERVOUS PEOPLE

Susanna Hertrich, Gesche Joost

ABSTRACT In this paper, we describe the Automatic Anchoring Armour, a wearable device synthesizing bio-feedback and mental conditioning. This synthesis is similiar to the technique used in Neuro Linguistic Programming (NLP): Anchoring. Following stress level analysis, AAA taps automatically and repeatedly on a specific spot of the forearm to trigger a positive emotional memory previously stored. Using the human skin as a receiver for haptic stimuli and interface to trigger positive emotions, the device is intended to change the wearer's perception of a social situation as being unsettling to being pleasant. This results in a feeling of calm.

The premise for using automation as a form of instant therapy in a wearable device with a bold presence is to stimulate a debate on the issue of cultivation of vulnerability raised by Furedi. This project explores issues of mental well-being and posts a critique of present-day therapy culture.

INTRODUCTION AAA is a device worn as protective armour for nervous people to cope with anxieties arising in social situations. The functionality of the device is modeled after the theory of Anchoring, a method of mental conditioning which forms part of the NLP repertoire of mental self-enhancement techniques.[1]

Following the logic of straight stimulus-response conditioning, Anchoring seeks to link a specific haptic sensation with a certain positive memory. In doing so, it seeks to evoke the positive feelings on-demand. The mental connection between the feelings associated with the memory and the haptic stimulus itself is in advance obtained through training. The above-described method is connected to a biofeedback loop, which senses the wearer's state of insecurity through a galvanic skin response (GSR) sensor[2]

DESCRIPTION AND USE The device comprises of a GSR sensor, a linear actuator, an Arduino board, a battery pack and other electronic components, embedded into a mounting. The mounting itself is attached to a leather wristband that holds the apparatus in place. The device is tightly

Fig. 1. Emotional pacemaker for therapy-seeking people.

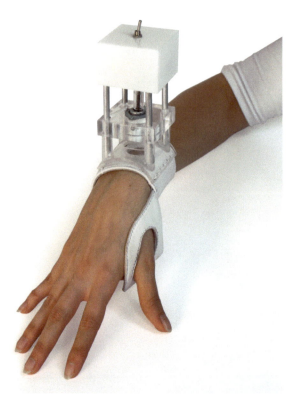

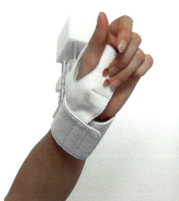

Top: Fig. 2 Using the human skin as receiver for haptic stimul..
Right: Fig. 3 GSR-sensor on the palm measures current stress levels.

strapped to the wearer's left lower forearm, with the thumb resting in a noose. To use the device, the wearer places her fingertips on the GSR sensor on the palm of her hand, which measures her current stress levels. If the level exceeds a certain threshold, the pressure function of AAA comes into action: The device performs pressure actuation which trigger the stored positive memory.

MACHINE CONDITIONING It is vital for the Anchoring technique to be executed at the same locus on the skin, with the same intensity. Since machines perform mechanical processes with calculated precision, and unlike human beings, remain undisturbed by emotion, they become an obvious choice of tool for this kind conditioning.

THERAPY CULTURE Furedi's term Therapy culture characterises the recent therapeutic turn towards emotional life, which widely replaces a focus on greater social interdependencies, resulting into the cultivation of vulnerabilities. Furedi argues, 'The therapeutic imperative is not so much towards the realisation of self-fulfillment as the promotion of self-limitation. It posits the self in distinctly fragile and feeble form and insists that the management of life requires the continuous intervention of psychological exercise.'[3]

CONCLUSION The goal of AAA is to take this point of critique to a physical manifestation in form of a device that acts as an emotional pacemaker for therapy-seeking people. As a design for debate[4] its exaggerated physical shape not only enable the device's functionality, it also amplifies the notion of AAA as protective equipment. Protective gear foretell the danger and 'project intimidating signs of superiority, whether technical, spiritual, or physical' [5] to the opponent. This way, AAA acts as a statement piece. It communicates the wearer's vulnerability while showing off a delusion of being under therapeutic guard.

Susanna Hertrich, Gesche Joost

References [1]. Bandler, R. and Grinder, J. (1979). Frogs into Princes: Neuro Linguistic Programming: Introduction to Neurolinguistic Programming. Real People Press, Boulder. [2] Shi, Y., Ruiz, N, Taib, R., Choi, E., and Chen, F. (2007). Galvanic skin response (gsr) as an index of cognitive load. In CHI '07: CHI '07 extended abstracts on Human factors in computing systems, pages 2651-2656, New York, NY, USA. ACM. [3] Furedi, F. (2003). Therapy Culture: Cultivating Vulnerability in an Uncertain Age. Routledge Oxford, 1st edition. [4] Dunne, A. and Raby, F. (2001). Design Noir: The Secret Life of Electronic Objects. Birkhäuser Basel, 1st edition. [5] Antonelli, P, Lowry G. D., O'Mahony, M., Patton P. and Yelavich, S. (2005). Safe: Design Takes On Risk. The Museum of Modern Art, New York.

SYNTHETIC EMPATHY: SOMAES-THETIC BODY ACTUATION AS A MEANS OF EMOTIONAL EVOCATION

Susanna Hertrich, Fabian Hemmert, Ulrike Gollner, Matthias Löwe, Anne Wohlauf and Gesche Joost

ABSTRACT In this paper, we present a novel means of emotional evocation: somaesthetic body actuation. This project explores ways to potentially evoke empathy and commiseration, even in times of supersaturated sadness. Three principles of emotional simulation through bodily means are discussed – fear, through coldness; panic, through constraint; and grief, through lachrymatory excitation. We propose three prototypes for this type of arousal: a hacked ice spray dispenser, a motorized contracting rig, and an onion-filled amulet.

INTRODUCTION Bad news headlines seem to be omnipresent in our everyday-lives. Be it for actual worsening of conditions in the world, or for mainstream media sensationalism: an increase of negative information can be observed across different everyday media. It may be hypothesized that this leads to a supersaturation of sadness. This supersaturation may, in turn, deaden the usually connected feelings of empathy and commiseration.

The imbalance of seemingly increasing bad news, but decreasing commiseration, which stands in contrast to the general human need for empathy. Providing the bodily signs of commiseration through synthetic means alters the perception of bodily reactions to societal circumstances – and their relation to the artificial.

The project underlines how design can help us to understand societal issues from a critical point of view, in our case, through a design noir[1] perspective that alters the perception of an emotional state through a change in its cause. No longer the mind causes the feeling, but a machine. The emotions of concern here are unpleasant, and generally suppressed. Yet, they are considered highly important – not only from an evolutionary point of view, but also from a socio-psychological one. According to Maslow, friendship is a basic human need, which may involve empathy and commiseration[2]. It appears relevant that bodily activity may have an effect on emotional states, as facial expression and body posture recently have proposed to do[3] – a finding that has been consistent across different cultures[4].

BACKGROUND In this section, we will review a number of projects that are related to the proposal at hand – we will analyze research endeavours in different areas and point to their potentials and shortcomings.

EMOTIONAL PHYSICAL COMPUTING Affect in human-computer interaction has been an active field of research in the past. Works by Picard[5] have pioneered the inclusion of emotion and affect into working with computers, and more recently been argued for by Boehner et al.[6]. Projects taking advantage of emotion include SenToy, a puppet posing-based input scheme[7], techniques based on eyebrow movements[8], and various others: Emotions are considered helpful for interacting with technology, but they are generally explored in how they can be used as a means of physical input, not in their evocation, as a physical output.

AWARENESS ON THE GO A variety of projects has explored the usage of context-awareness in mobile devices[9] to provide new experience to users, through music players[10], historical information[11], tourist guides[12] and sporting activities[13]. Other applications include belts and headbands that convey directional[14] and distance information[15] – the spectrum of applications.

INVADING THE BODY SPACE Explorations into the realm of inside the body are an objective of a large body of research. Such endeavours involve brain stimulation[16], nanowires[17] and body site-targeted drug injections[18]. Given their immediate impact on bodily being, such technology can provide intense experiences, but it may, at the same time, be not researched adequately in how much non-invasive actuation on the body's surface could help to provide help in emotional evocation. Given general skepticism against body-invasive technology, it may be worthwhile to leverage on the perceptive characteristics of the body's surface.

The areas of research into emotion, awareness and body space invasion that have been presented in this section offer rich potential: they can be helpful in task fulfillment, serve flexibly in a broad spectrum of applications, and also provide rich, intense moments of interaction. At the same time, they have shortcomings in the balance between physical input and output, in their emotional richness, and in the fact that they are often bound to invasive implementations. We propose to assess this issue through an approach of bodily actuation that is emotionally evocative,

Susanna Hertrich, Fabian Hemmert, Ulrike Gollner, Matthias Löwe, Anne Wohlauf and Gesche Joost

Fig. 1: Fear simulation, back-worn ice spray actuation.

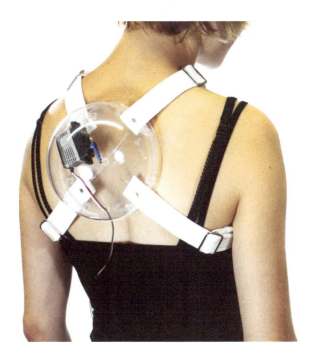

Fig. 2: Panic simulation, torso-worn contractor actuation.

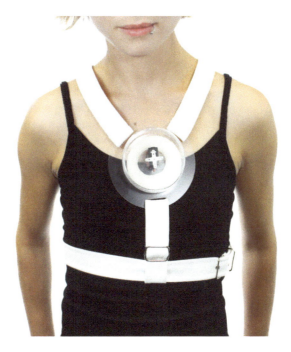

Fig. 3: Grief simulation, neck-worn onion actuation.

but yet available as a solution that operates on the body's surface. We propose to do so in a low-tech approach, a technical solution is meant that is created from off the-shelf materials, by simply rearranging them in unconventional ways.

Susanna Hertrich, Fabian Hemmert, Ulrike Gollner, Matthias Löwe, Anne Wohlauf and Gesche Joost

PROTOTYPES We present a set of three prototypes, allowing for the evocation of bodily experiences that are usually connected to emotional reactions bad news. The proposed emotions are chosen to relate to an unpleasant event in the future, present, and past, in the order they would normally occur: coldness, as an experience of fear; constraint, as an experience of panic; and lachrymatory excitation, as an experience of grief. All prototypes are controlled through a nearby Arduino board. The board is connected to a mobile phone through a computer, allowing it to read live data from the internet, which are then represented in an actuation.

We present a device that, worn on the back, allows for the evocation of shivers through the release of an ice spray on the user's back (Fig 1). The device proposed offers a low-tech approach to create a creeps-like feeling, based on the assumption that the skin reacts to a rapid cooling in a similar way like shivering. The ice spray is mounted in a rig that is attached to the wearer's shoulders and stomach, the cap is pressed by a magnetic coil.

The second device we propose allows for an artificial contraction of the torso (Fig 2), simulating a panic-like constraint of breathing. It does so in contracting a rig which is worn around the upper body. The rig can be contracted either abruptly or gradually, which may result in different experiences.

The third device is a lachrymatory actuation system, causing tears in the users eyes. This is achieved through an amulet which contains a freshly cut onion. A built-in motor moves the amulet's upside, opening a hole in its surface that releases the onion's scent.

CONCLUSION We have presented a set of prototypes for novel ways of bodily actuation. The presented devices are of bricolage-like character, demonstrating how low-tech prototyping can help us to explore new ways of interaction. In how much these ways can are not only bodily, but also actually emotionally evocative has to be determined in future studies. While providing insights into possible new ways of interaction seemed to have been a worthwhile goal on the one hand, it appears that the presented devices, standing by themselves, may be a valuable basis to stimulate a debate on recent societal developments. So much for the good news.

References [1] Dunne, A., & Raby, F. (2001). Design Noir: The Secret Life of Electronic Objects. Birkhäuser Basel. [2] Maslow, A. (1987). Motivation and Personality. HarperCollins Publishers. [3] Schnall, S., & Laird, J. (2003). Keep smiling: Enduring effects of facial expressions and postures on emotional experience and memory. Cognition & Emotion, 17(5), 787-797. [4] Ekman, P. (2007). Emotions Revealed, Second Edition: Recognizing Faces and Feelings to Improve Communication and Emotional Life. Holt Paperbacks. [5] Picard, R. (2000). Affective Computing. The MIT Press. [6] Boehner, K., DePaula, R. r., Dourish, P., & Sengers, P. (2005). Affect: from information to interaction. Paper presented at the CC '05: Proceedings of the 4th decennial conference on Critical computing, Aarhus, Denmark. [7] Vala, M., Paiva, A., & Prada, R. (2004). From Motion Control to Emotion Influence: Controlling Autonomous Synthetic Characters in a Computer Game. Paper presented at the AAMAS '04: Proceedings of the Third International Joint Conference on Autonomous Agents and Multiagent Systems, New York, New York. [8] Surakka, V., Illi, M., & Isokoski, P. (2004). Gazing and frowning as a new human-computer interaction technique. ACM Trans. Appl. Percept., 1(1), 40-56. [9] Häkkilä, J., Schmidt, A., Mäntyjärvi, J., Sahami, A., Aakerman, P., & Dey, A. (2009). Context-Aware Mobile Media and Social Networks. Paper presented at the MobileHCI '09: Proceedings of the 11th International Conference on Human-Computer Interaction with Mobile Devices and Services, Bonn, Germany. [10] Reddy, S., & Mascia, J. (2006). Lifetrak: music in tune with your life. Paper presented at the HCM '06: Proceedings of the 1st ACM international workshop on Human-centered multimedia, Santa Barbara, California, USA. [11] Ballagas, R., Kratz, S., Borchers, J., Yu, E., Walz, S., Fuhr, C., et al. (2007). REXplorer: a mobile, pervasive spell-casting game for tourists. Paper presented at the CHI '07: CHI '07 extended abstracts on Human factors in computing systems, San Jose, CA, USA. [12] Abowd, G., Atkeson, C., Hong, J., Long, S., Kooper, R., & Pinkerton, M. (1997). Cyberguide: a mobile context-aware tour guide. Wireless Networks, 3(5), 421-433. [13] Mueller, F., O'Brien, S., & Thorogood, A. (2007). Jogging over a distance: supporting a "jogging together" experience although being apart. Paper presented at the CHI '07: CHI '07 extended abstracts on Human factors in computing systems, San Jose, CA, USA. [14] Nagel, S., Carl, C., Kringe, T., Märtin, R., & König, P. (2005). Beyond sensory substitution: learning the sixth sense. Journal of Neural Engineering, 2(4), R13. [15] Cassinelli, A., & Reynolds, C. (2006). Augmenting spatial awareness with Haptic Radar. Paper presented at the In: Proceedings of the 10th IEEE International Symposium on Wearable Computers. [16] Rosenberg, O., Zangen, A., Stryjer, R., Kotler, M., & Dannon, P. (2009). Response to deep TMS in depressive patients with previous electroconvulsive treatment. Brain Stimulation. [17] Wang, X., Song, J., Liu, J., & Wang, Z. (2007). Direct-Current Nanogenerator Driven by Ultrasonic Waves. Science, 316(5821), 102-105. [18] Ghanem, A., Steingen, C., Brenig, F., Funcke, F., Bai, Z.-Y., Hall, C., et al. (2009). Focused ultrasound-induced stimulation of microbubbles augments site-targeted engraftment of mesenchymal stem cells after acute myocardial infarction. Journal of Molecular and Cellular Cardiology, 47(3), 411-418.

Susanna Hertrich, Fabian Hemmert, Ulrike Gollner, Matthias Löwe, Anne Wohlauf and Gesche Joost

BODY AS AN INTERFACE. EXPERIENCING OF PHENOMENA IN AN INTERACTIVE ART

ABSTRACT The connection of digital technology and art in 20th century has brought many new questions into the field of aesthetics. Almost 50 years of the close co-existence of electronic devices and artistic practice continuously forces theoreticians to re-think their subject, in the name of involving very unconventional ideas of using new electronic media as a basis for many contemporary artistic works. Philosophers try to expand the aesthetic theory, and give the most appropriate perspectives to value creative technology used by artists. They also test previous aesthetic tendencies in the light of new born aesthetic terms and problems.

In this paper I would like to present a phenomenological approach to aesthetics, called Aesthetics of Appearing, introduced by German philosopher Martin Seel, and compare his vision of the fresh aesthetic view with examples of new media art: digital environment based interactive installations, and interactive performances – works based on untypical interfaces, which make a use of human body and his percepts.

In his Aesthetics of Appearing[1], Martin Seel continues the philosophical tradition of German phenomenological school, and suggests the new approach into aesthetics – the first aesthetical project based on phenomenology. The main notion used in this project is to appear, which corresponds to treating the aesthetic experience as experiencing of things and events exactly in the way they appear to our senses in a particular moment, to register them simultaneously - as it was the only natural way of meeting the outside world for a human being. This special kind of attention of the viewer – the participation in the world which he is observing – is actually the attention given to himself: focusing on appearing of the outside world by combining the images with personal experiencing structures, gives us back two kinds of information: one is about the outside reality (material one, and ideal one – depends on the level of reflection), and second is about us existing on these levels of the outside world. Seel stress that the artworks are particularly interesting to investigate,

Katarzyna Otulakowska

because they generate probable and improbable presents, and they speak to all of human senses – that is how they use the energy of their own existence (without the need of being oriented on any practical result).

Moreover, concentrating on appearing of works of art in a particular moment, is also seen as directing our attention to situation of perception – a strong orientation of oneself on using the percepts in a light of getting a feedback information from the reality/situation one exists in. Finally, that is the way of getting a self-knowledge on many cognitional levels.

What is more, Seel stresses that every aesthetic cognition is based on the whole sensorium – there is no possibility to use one sense only in experiencing a work of art. He claims that the idea of synesthesis is one of these which we should come back to: contemporary artworks visibly attack more than just one main sense, and provoke more reflection dedicated to the construct of human perception.

At this basis, it is important to examine some of the examples of digital art – one of the most important field of today's artistic practice. Repeating by Manovich, we can see that 'the picture, in a traditional meaning of this word, does not exist anymore', and 'the screen has disappeared it has completely filled up our field of vision'[2], as a consequence of interactivity. Therefore, we would like to follow some examples of interactive installations and performances, and try to investigate the problem of a human body used as an interface, which allows us to see the appearing of these digitally programmed artistic phenomena. We do treat interactivity as a paradigm of digital art, and we believe it changes the philosophical thought about human condition. We would like to examine following works:

(1) Lynn Hershman, Room of One's Own (1993) – interactive installation

(2) Christa Sommererand Laurent Mignonneau, A-Volve (1994) – Virtual reality

(3) Stelarc, Movatar – interactive performance

(4) Paul Sermon, Telematic Dreaming (1992) – interactive installation

(5) MarekChołoniewski, Darken Light Sound (1999) – interactiveperformance

All of these examples focus on different use of human body – either it is the body of the artist, or bodies of viewers. Every one of these works build different environments of experiences, and all of them bring questions about human condition since its meeting with the technological interfaces.

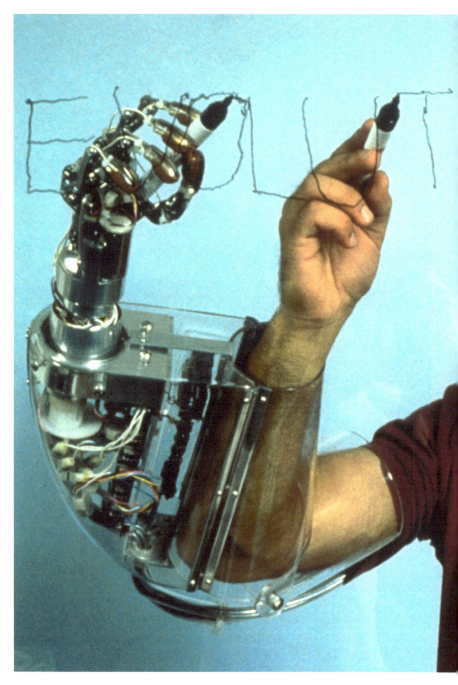

Stelarc, Handwriting – Evolution: Writing One Word Simultaneously With Three Hands
Maki Gallery Gallery, Tokyo, 1982. Photographer:Keisuke Oki

28

Interaction in digital art provokes to use human body as an interface – we see the examples of treating the body as two kinds of interface: a perfect one, whole connected with the artificial artistic reality (immersion), and a half-perfect, which allows us to travel between the levels of virtual and real worlds (as communication between them). 'The work of art is not an image anymore, is not a twodimensional window to the world, but becomes a portal to the multisensual event. The observer is both outside and inside of this event, becoming a part of what he's observing'[3].

Katarzyna Otulakowska

References [1] Seel M., Aesthetics of Appearing. [2] Manovich L., The Language of New Media. Wydawnictwa Akademickie i Profesjonalne, Warszawa, 2006, 129-135. [3] Weibel P., Ars Electronica. An Interview by Johan Pijnapel. Art & Design vol. 9, nr 11-12, 26-31.

SURFEL CAMERAS

Depth-camera systems provide a representation of visual scenes that combine an image with depth coordinates. These camera systems, which use principles like stereo disparity and time of flight, take samples of the surface of objects. It is convenient to model the imagery produced by depth-cameras using surfels[1] as opposed to polygons.

Many interaction techniques exist for polygons and 2D imagery, but few have been proposed explicitly for surfels. As surfels are not explicitly connected in the same manner as polygons, interaction with them can take the form of set or point operations. Adams and Dutré[2] previously showed how boolean operations could be implemented on solids bounded by surfels. Kim and Choi[3] further apply surfels in creating a system to simulated application of facial makeup.

Surfel cameras fuse information normally captured separately using motion capture and texture capture systems. Unlike tomography, surfel cameras do not provide 3D imagery of the interior of objects. Interestingly, surfel cameras produce objects that can be interacted using operators more often found in 3D modeling systems than in photographic manipulation systems. Pointshop 3D[4] provides one instance of a system for editing surfel data.

IMPLEMENTING A SURFEL CAMERA Inspired by the work of Gluckman et al.[5], myself and others set out to develop a high-speed surfel camera system. Making use of a pair of 200 frame per second firewire cameras equipped with omnidirectional mirrors, the oversights system seeks to capture 360-degree depth images. Through the CUDA toolkit, the system uses a graphics accelerator to improve the performance of distortion removal and stereo matching processing programs.

The raw images captured from the omnidirectional mirrors are distorted. A camera model captured using Mei's omnidirectional calibration toolbox[6] is used to provide de-warped panorama images. For each of the stereo cameras, the original image is warped into a panorama image which spans 360 degrees.

The two panoramas are in turn fed into a stereo-disparity system to compute the depth coordinates for each pixel. Kohei Yamaguchi adapted some code written by David Gallup et al.[7] to perform stereo matching using the

Graphics Processing Unit (GPU). The approach used is to search along the epipolar line between the images using sum of square differences. We found that this system is able to compute surfel images at a rate exceeding 150 Hz. By way of comparison, a system using similar algorithms but making use of a dual-core CPU operates at approximately 0.5 Hz.

The image processing is embarrassingly parallel as much of the computation takes the form of pixel operations with no data interdependencies. By taking a General Purpose GPU[8] approach to the problem, we are able to use tens of thousands of threads to processes each omnidirectional frame. The floating point and arithmetic operation through-put offered by the comparatively larger die sizes of modern GPUs such as the Nvidia GTX 480 mean that surfel cameras can be quite fast. Moreover, surfel cameras can be made by coupling commodity GPU systems with increasingly inexpensive high-frame-rate cameras.

The panorama imagery and depth map are then fed into an OpenGL based rendering system developed by Elias Freider that displays the surfels. From one perspective, the viewer sees what appears to be a typical panorama. However, under rotation one can examine the surface manifold from different perspectives giving the sensation that apes viewing an object from different 3D viewpoints.

SURFEL MOTES The combination of a surfel cameras with wireless data links allows for an unusual multi-camera system. Typically, a surfel camera only views one perspective of an object. In practice this means that a surfel camera looks as if it is capturing a mold of one face of an object. A network of surfel cameras could view an object from many different but overlapping perspectives. Moreover, if the position of the surfel cameras is precisely known, then it is possible to infer the location and texture of objects in a scene in real time.

Extending of the work of Carceroni and Kutulakos[9] we can easily imagine a world in which arbitrarily placed surfel cameras communicating with one another are able to capture the texture and motion of an entire room's worth of contents. Instead of reviewing video, a user could replay the evolution of 3D surfaces as they moved through space. Such systems would be of great use to game designers as well as those working in 3D animation.

This paper has served as a brief review of interrelated work from communities researching surfels, camera systems, and interactive graphics. The

33

oversights system and yet to be developed surfel camera systems allow a different perspective of scenes combining motion capture with surface texture information.

ACKNOWLEDGEMENTS The oversights system was developed in collaboration with Kohei Yamaguchi and Elias Freider. Yoshihiro Watanabe and Hiromasa Oku also provided valuable suggestions for the mounting and camera calibration of the oversights system.

Carson Reynolds

References [1] Pfister, H., Zwicker, M., Baar, J. van, & Gross, M. (2000). Surfels: surface elements as rendering primitives. International Conference on Computer Graphics and Interactive Techniques, 335. Retrieved from http://portal.acm.org/citation. cfm?id=344779.344936 [2] Adams, B., & Dutré, P. (2003). Interactive boolean operations on surfel-bounded solids. ACM Transactions on Graphics (TOG), 22(3), 651. Retrieved from http://portal.acm.org/citation.cfm?id=882262.882320. [3] Kim, J.-S., & Choi, S. (2007). A virtual environment for 3D facial makeup. Lecture Notes In Computer Science, 488-496. Retrieved from http://portal.acm.org/ citation.cfm?id=1770146. [4] Zwicker, M., Pauly, M., Knoll, O., & Gross, M. (2002). Pointshop 3D: an interactive system for point-based surface editing. International Conference on Computer Graphics and Interactive Techniques, 21(3), 322. Retrieved from http://portal. acm.org/citation.cfm?id=566584 [5] Gluckman, J., Nayar, S. K., & Thoresz, K. J. (1998). Real-time omnidirectional and panoramic stereo. In: Proceedings of the 1998 DARPA Image Understanding Workshop (Vol. 1, pp. 299-303). Retrieved from http://citeseerx. ist.psu.edu/viewdoc/summary?doi=10.1.1.29.6913 [6] Mei, C. (2007). Omnidirectional Calibration Toolbox. Retrieved from http:// www.robots.ox.ac.uk/~cmei/Toolbox.html. [7] Gallup, D., Frahm, J., & Stam, J. (2009). CUDA Stereo. Retrieved from http://www.cs.unc. edu/gallup/cuda-stereo/. [8] Luebke, D., Harris, M., Krüger, J., Purcell, T., Govindaraju, N., Buck, I., et al. (2004). GPGPU: general purpose computation on graphics hardware. SIGGRAPH 2004: ACM SIGGRAPH 2004 Course Notes (p. 33+). New York, NY, USA: ACM. doi: 10.1145/1103900.1103933. [9] Carceroni, R. L., & Kutulakos, K. N. (2002). Multi-View Scene Capture by Surfel Sampling: From Video Streams to Non-Rigid 3D Motion, Shape and Reflectance. International Journal of Computer Vision, 49(2), 175. Retrieved from http:// portal.acm.org/citation.cfm?id=598521/.

DEVICES THAT ALTER....
A POTTED INQUIRY

According to Rhodes et al., at the MIT Wearable Computing Project, 'in 1268 Roger Bacon made the first recorded comment on the use of lenses for optical purposes. However, by that time reading glasses made out of transparent quartz or beryl were already in use in both China and Europe'[1]. A short four hundred years later, in 1665, Robert Hooke references eyeglasses when he calls for augmented senses in his preface to Micrographia (Hooke, 1665, in Rhodes, et al): 'The next care to be taken, in respect of the Senses, is a supplying of their infirmities with Instruments, and as it were, the adding of artificial Organs to the natural ... and as Glasses have highly promoted our seeing, so 'tis not improbable, but that there may be found many mechanical inventions to improve our other senses of hearing, smelling, tasting, and touching.'

Today, another four hundred years or so after Hooke, we are still on the path of sensory augmentation for the purpose of altering perception. But the bar has shifted somewhat from bringing under-functioning percepts to body typical performance levels, to bringing cyborg aesthetics and transhumanist concerns[2,3] into everyday life. If we step away from the extremes proposed and provided by Stelarc[L1], Orlan[L2] and the more radical members of the body modification community,[L3] we find hearing-impaired people from all walks of life with cochlear implants that stimulate nerves to provide the perception of sound. We also find Paralympic champions whose eligibility to compete in Olympic trials are hotly contested as they are perceived to be advantaged by their prosthetic attachments[L4], and the American military pushing for the extremes suggested by Hooke in 1665 to be accepted as standard procedure[L5]. In addition, in living rooms today, families can play EEG/mind control games that challenge our perception and understanding of what we thought were ubiquitous physical laws, as with thought alone a player may make a ball levitate (by controlling the speed of a fan which gives it lift)[L6] or undertake other activities, depending on how the interface has been (re)purposed[L7]. Our perception of the norm, and of body-typical, with regard our percepts, seems to be transmuting.

Danielle Wilde

Historically, head-mounted displays, gloves and other percept altering devices, have come from the arts rather than the technical community [4]. Ivan Sutherland's pioneering work introducing head-mounted displays to virtual reality[5], for example, developed around the same time as Lygia Clark's Sensorial Hoods, (1967)[6], Walter Pichler's TV-Helmet (Portable Living Room)[s] and Small Room prototypes, (1967)[7], and Haus-Rucker-Co's Environmental Transformers[L8] (1967 and 68). The visual and cultural parallels between these works are significant and arguably unexplored[6]. By looking to the arts and conceptual and propositional design we can find a bevy of more or less wild, entertaining, stimulating, provocative proposals aimed at prompting us to reflect on the future we would like to live in, as well as how we position ourselves in relation to current society, accepted norms, and the role of technology in our lives today. Dunne and Raby[8,L9] and the bevy of students and researchers who have followed their lead at the Royal College of Art[L10] and elsewhere provide some provocative examples.

Outside the hallowed realms of conceptual and propositional art and design, we find the proposals of the controversial cosmetic surgeon, Dr Joseph Rosen, Professor of Surgery at Dartmouth Medical School[L11] and Adjunct Professor of Engineering at Thayer[L12]. Dr Rosen stated in 2001 that within five years he would be able to give people wings[L13]. When we have a limb amputated, our neural map of that limb gradually fades away; if we gain a body part, our neural map expands accordingly. The wings could therefore be incorporated into an individual's understanding of themselves.

'If I were to give you wings, you would develop, literally, a winged brain. Our bodies change our brains, and our brains are infinitely mouldable', according to Rosen[9]. This begs the question: does our neural map expand when we wear temporary devices that alter perception? And: how much wear or use is required to shift what our brains consider to be the norm?

Dr. Rosen questions the conservative medical restraints that prevent him from deploying some of his most creative visions: wings for humans; cochlear implants to enhance hearing to the range of an owl; rods to provide binocular vision, to see for miles and into depths as well. He has ideas for implants, gadgets, gears, discs, buttons, including sculpting soft cartilage to enable us, as humans, to cross the frontiers of our own flesh and emerge as... as what[9]? Rosen is an advisor to NASA, and has, with

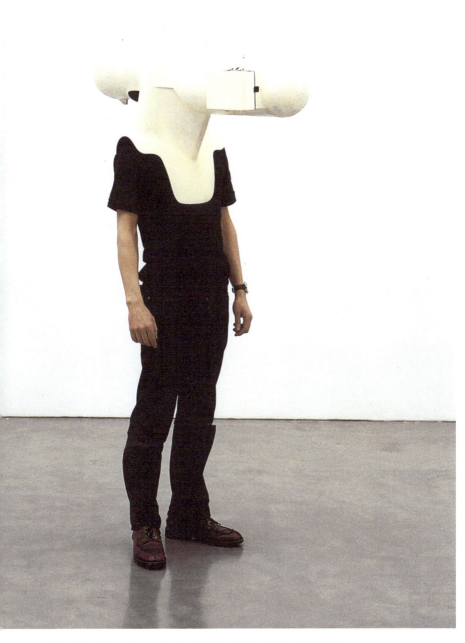

Walter Pichler TV-Helmet (Portable Living Room), 1967 © Jean Tinguely Museum in Switzerland

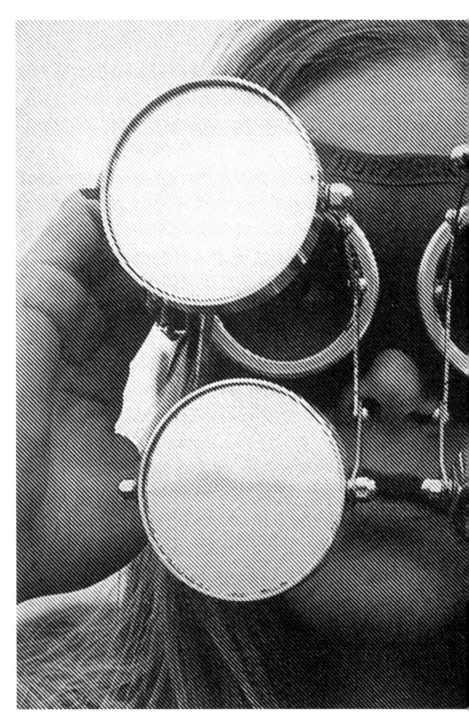

Top: Lygia Clark Eyeglasses, 1968 © 2006 Artists Rights Society (ARS), New York / ADAGP, Paris
Top right: Lygia Clark Sensory Masks, 1967 © 2006 Artists Rights Society (ARS), New York / ADAGP, Paris

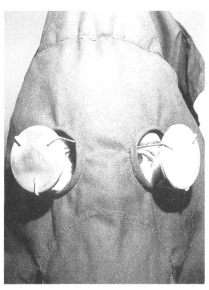

Danielle Wilde

his colleagues, expanded the boundaries of his surgical specialty and medicine in general by looking at the world around him and asking why. Most people, according to his University[L14], missed Rosen's real message in 2001: that plastic surgeons solve problems that don't have solutions. I ask, is this not the role of designers?

why is often asked when people are confronted with wearable devices that alter our percepts, but that why may come from a different perspective than Rosen's. On design boom's web post on wearable computing, for example, the first comment essentially asks why the proposed perception altering device is even needed – why is it seen as an enhancement, or an advantage over life without it? The other comments reiterate this question[L15]. So why do we continue along this pathway of creating devices that alter perception? What drives the artists and designers, engineers and tinkerers? What use can the devices possibly be?

Shanon Larratt, the creator and former editor and publisher of BMEzine[L16], the oldest and largest body modification website on the internet asks:

Do we want something that's going to be 'neat' for 15 minutes, or something that will permanently enrich our lives?[L17] I extend this to ask: are propositional devices enough to raise provocative questions? Or do we need to make the objects and experiences being proposed? What is the difference, for example, between Auger-Loizeau's Audio Tooth Implant[L18] and Stelarc's planned realization of a mouth-mounted USB cell-phone receiver[L19]? Are developers even engaging with these questions?

IDEO's Fred Dust, speaking at TedxBerkeley in April this year[L20] gives examples of how good design doesn't just look beautiful, it acts differently and it makes the people who use it act differently. It would seem that any device that alters perception (DAP) would prompt a person who is wearing or otherwise engaging with it to act differently. How can we discern, then, what makes a DAP well designed? Dust provides us with three principles he sees as essential to good design:

(1)   Know the constraints within which you are designing.

(2)   Be empathic – think seriously about the way you feel and the way that people around you might feel, and use that as inspiration for design.

(3)   Challenge precedence, break through the way things have been done,

(4)   and prototype your idea – try it, put it out there, test it.

How do these principles map onto devices that alter perception? What kind of criteria can we employ to help us, as a community, push forward beyond an ever-growing, rich, yet straggled collection of ideas to a coherent forward thinking, and evolving body of knowledge?

Pranav Mistry's Sixth Sense give[s] you relevant information about whatever is in front of you[L21]. It alters our sense of agency by giving access to information anywhere, any time through natural gesture[L22]. It has been touted as the first wearable that seems useful[L23].

The questions I pose are these:

→ What constitutes usefulness in the context of DAPs?

→ Is design for debate, in relation to DAPs, effective, and thereby (or otherwise) useful?

→ What defines a DAP?

→ How do we evaluate such devices?

→ How can we bring rigorous methodologies to their development, evaluation and distribution, (whether the final embodiments are prototypes, products or scenarios)?

→ How can designers, artists, theorists, makers, engineers and technologists cross-fertilize in meaningful ways, and thereby enrich their enquiry,

and ultimately the DAPs they are developing?

Danielle Wilde

References [1] Rhodes, B., Starner, T., Maguire, C., Platt, D., Pentland, S., Urban, R., Rekimoto, J., Matias, E., Becker A. and others. Online: http://www.media.mit.edu/wearables/lizzy/timeline.html#1268. [2] Harraway, D. A cyborg manifesto: Science, technology and socialist-feminism in the late twentieth century. In: Symians, Cyborgs and Women: The Reinvention of Nature Free-Association Books, London, 1991. pp.149-181. [3] Bostrom, N. A History of Transhumanist Thought. In Journal of Evolution and Technology 14,1 ⊟ April 2005. [4] Krueger, M.W. The Artistic Origins of Virtual Reality. SIGGRAPH Visual Proceedings ACM, New York, 1993 pp. 148-149. [5] Sutherland, I.E. A head-mounted three dimensional display. Proc. AFIPS Joint Computer Conferences, Part 1. San Francisco, CA. December 1968. ACM, New York. pp. 757-764. [6] Osthoff, S. Lygia Clark and Hélio Oiticica: A Legacy of Interactivity and Participation for a Telematic Future. In: A Radical Intervention: The Brazilian Contribution to the International Electronic Art Movement, Ed. Eduardo Kac. Leonardo 30, 4. MIT Press, 1997. [7] Breitwieser, S. Pichler: Prototypen/prototypes 1966-69. Generali Foundation, Vienna, 1998. [8] Dunne, A., Raby, F. Design Noir: The Secret Life of Electronic Objects, Birkhäuser Basel. 2001. [9] Slater, L. DR. DAEDALUS (eccentric plastic surgeon Joe Rosen)(Interview) Harper's Magazine 2001. [10] Zweite, A., Schmidt, K., von Drathen, D. and Horn, R. Rebecca Horn, Drawings, Sculptures, Installations, Films 1964-2006. Hatje Cantz, Berlin, 2007. Links [L1]http://v2.stelarc.org/projects.html [L2] http://www.orlan.net/ [L3] http://www.BMEzine.com/ [L4] http://www.nytimes.com/2007/05/15/sports/othersports/15runner.html?_r=2&oref=slogin—http://www.iaaf.org/news/kind=100/newsid=42384.html [L5]http://tacticalwarfightergear.com/tacticalgear/catalog/soldier_exoskeleton.php—http://tacticalwarfightergear.com/tacticalgear/catalog/nano_technology_military.php—http://tacticalwarfightergear.com/tacticalgear/catalog/Cyborg_Soldiers.php [L6] http://mindflexgames.com/how_does_it_work.php [L7]http://www.neurosky.com/—http://www.engadget.com/2010/09/07/apples-in-stereo-man-controls-his-moog-with-his-mind-and-yo/—http://www.engadget.com/2010/03/04/mattel-mindflex-hack-shocks-you-into-serenity/—http://www.engadget.com/2009/01/13/star-wars-force-trainer-teaches-children-to-kill-with-their-m/-http://ericmika.com/itp/brain-hack [L8]http://anthropomorphe.blogspot.com/2009/05/second-nature.html [L9]http://www.dunneandraby.co.uk [L10]http://www.interaction.rca.ac.uk/—and—http://www.designproductsrca.com/ [L11]http://dms.dartmouth.edu/ [L12]http://engineering.dartmouth.edu/ [L13] http://www.dartmouthengineer.com/2009/09/doctor-in-the-class/ [L14] http://www.dartmouthengineer.com/2009/09/doctor-in-the-class/ [L15]http://www.designboom.com/weblog/cat/16/view/5586/wearable-computers.html [L16]http://www.BMEzine.com/ [L17]cited online at http://www.cyberpunks.org [L18]http://www.auger-loizeau.com/index.php?id=7 [L19]http://v2.stelarc.org/projects/earonarm/index.html [L20]http://tedxberkeley.org/media/fred-dust [L21]http://www.ted.com/talks/lang/eng/pattie_maes_demos_the_sixth_sense.html [L22]dependent on a web connection [L23] http://www.styleborg.com/

41

EMPATHY MIRRORS

ABSTRACT   We describe two objects motivated by a theory of empathy. The first, the Tear Mirror causes artificial or collected tears to roll down the cheeks of a recipient of an expression from a crying individual. The second, the Countenance Mirror produces haptic stimulation conveyed by air puffs on the part of the face of a recipient linked to an emoting individual.

The mirror neuron system is believed to be critical for interpreting facial emotions[1]. By building devices whose design is motivated by the mirror neurons, we set out to entrain the emotional experiences of communicating individuals. The Tear Mirror and Countenance Mirror are designed to provide an experience of empathy through the use of twin cues.

MOTIVATION   It has been hypothesized that affect mirroring plays an important role in emotional development. Likewise experimenters found[2] that emotional states such as happiness and disgust can be induced. In particular researchers observed emotion-specific autonomic nervous system activity when participants performed facial actions[3] that imitate expressions.

This tells us that emotions can be telegraphed from one individual to another. It is worth asking: why should emotions or empathy be enhanced? Without empathy, basing our interactions on purely rational behavior, many problems arise. One is that rational bargaining can lead to sub-optimal situations such as the prisoner's dilemma. Frank[4] argues that emotions and empathy let us escape from problems rooted in pure rationality. Damassio[5] found patients whose affective nervous system was damaged were not able to perform everyday tasks. Their unbounded rationality led to many unsatisfactory situations. Franz de Waal[6] argues that all mammal need empathy in order to survive.

In computer-mediated communication, we find that emotion is often masked or requires extra effort to express. Many of us have written emails with emotional content which were terribly misunderstood. Some systems, like emotemail[7] try to capture this extra emotional expression. However, these systems still rely on the recipient to interpret the expressed emotion. When trying to capture and communicate emotion many problems may exist. One is the analysis problem: does a device or system correctly identify and label an emotion being expressed. A second problem is the

Tomoko Hayashi and Carson Reynolds

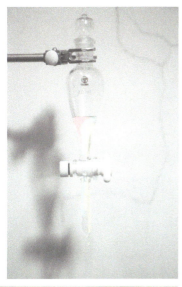

Tomoko Hayashi and Carson Reynolds

interpretation problem: does the device's representation of the emotion convey to the recipient what the sender is feeling?

Thinking about these problems from the perspective of devices that alter perception a different approach appears. That is to make a devices that causes the recipient to perceive the sender's emotional state directly skipping the analysis and interpretation problems.

HOW WE CAN BUILD EMPATHY MIRRORS? One approach to the tear mirror would be to collect a large number of children and make them cry. After bottling their tears, these valuable liquids could be re-heated and re-served. However this approach may be unethical.

Joking aside, the Tear Mirror would require a liquid which felt similar to actual tears especially in terms of saltiness and temperature. So one important project will be to create simulated tears. An important question would be if there is a palpable difference between happy tears and bitter tears. The Tear Mirror would also need to control the tear tap to regulate how many tears were being communicated: just teary eyed or distraught convulsive crying.

A second step would be to detect when a person is crying. A video system showing the person's face could look for tell-tale streaks of tears or twitches and facial muscle contractions that might occur when someone cries. An audio analysis system could also try to pick out the sound of a the sender's cries. A good starting point for an audio project would be to collect a corpus in which some speakers are crying while others are talking with more positive emotional states.

Putting together the tear detector and properly simulated tears would work together to emit streams of tears on the recipient's face perhaps through tubes that would empty near the corner of the eye or from the dripping tap above the specially designed chair.

The countenance mirror takes a more subtle approach by creating tingling on the recipient's face in the same area as the sender's facial muscles are moving. And so an arched eyebrow would causes air puffs to be directed above the recipient's eyebrow.

This would require a mapping between the communicating pair's faces. Since (with the exception of twins) most human being's faces have a distinct shape, the system would need to take into account these differences. Interestingly, such mappings might allow adults and children to communicate despite having faces of different sizes.

Tomoko Hayashi and Carson Reynolds

In terms of the actual stimulator we can imagine two possible designs. One would be like an LCD screen: a matrix of pixel puffers which would be activated to send out air puffs in particular parts of someone's face. A different idea would be to use a small number of air puff jets that could be aimed via pan and tilt. This would be more like a cathode-ray tube in that the jets would have to scan over the person's face.

Clearly we are still at the point of imaging empathy mirrors. After gathering some feedback from the workshop participants we'd like to start making simple prototypes of the empathy mirrors.

RELATED WORK Dobson[8] built a number of therapy devices that encouraged humans to express their emotions. Although her approach often focuses on human to machine interactions, these experiences of perceptual resonance can also be applied to human to human communication. Hertrich[9] has discussed devices that augment the instincts by enhancing emotions such as fear and anxiety. Although, these projects often focus on linking these emotions to news as sources of emotional experience. Vaucelle and Bonanni's haptic interfaces for mental therapy[10,11] focus not only the pleasurable touch but also painful touch which shows the potential of sharing pain to induce the feeling of closeness between two people. Auger's[12] pieces on scent and communication show one way of sharing emotion through the animalistic nature of smell. Manabe's pieces on stimulating the facial muscles[13] could be used to share emotions and induce the feeling of empathy through facial expressions.

Tomoko Hayashi and Carson Reynolds

Reference [1] Williams, J. (2001). Imitation, mirror neurons and autism. Neuroscience & Biobehavioral Reviews, 25(4), 287-295. doi: 10.1016/S0149-7634(01)00014-8. [2] Laird, J. D., & Strout, S. (2007). Emotional behaviors as emotional stimuli. Handbook of Emotion Elicitation and Assessment (Series in Affective Science). Oxford University Press, USA. Retrieved from http://www.worldcat.org/isbn/0195169158. [3] Ekman, P., Levenson, R. W., & Friesen, W. V. (2008). Autonomic Nervous System Activity Distinguishes among Emotions. Science, 221(4616), 1208-1210. [4] Frank, R. H. (1988). Passions Within Reason: The Strategic Role of the Emotions. Journal of Political Economy (Vol. 98, p. xxv, 368 p.). W W Norton & Co Inc. Retrieved from http://www.loc.gov/catdir/toc/ecip0815/2008012338.html. [5] Damasio, A. R. (1994). Descartes' Error: Emotion, Reason, and the Human Brain. New York Avon (Vol. 310, p. 0). Putnam. Retrieved from http://www.amazon.com/Descartes-Error-Emotion-Reason-Human/dp/014303622X. [6] Waal, F. D. (2009). The Age of Empathy: nature's lessons for a kinder society (p. 35). Harmony Books. [7] Angeslevä, J., Reynolds, C., & O'Modhrain, S. (2004). EmoteMail. In SIGGRAPH '04: ACM SIGGRAPH 2004 Posters. New York, NY, USA: ACM. Retrieved from citeulike-article-id:1902291. [8] Dobson, K. (2004). Blendie. Designing Interactive Systems, 309. Retrieved from http://portal.acm.org/citation.cfm?id=1013115.1013159. [9] Hertrich, S, & Reynolds, C. (2009). Prosthetics for the Instincts. In Proceedings of International Symposium on Electronic Art (ISEA 2009). Belfast, Northern Ireland. [10] Bonanni, L., & Vaucelle, C (2006). Affective TouchCasting Proceedings of International Conference on Computer Graphics and Interactive Techniques. Retrieved from http://portal.acm.org/citation.cfm?id=1179849.1179893. [11] Bonanni, L., & Vaucelle, C. (2009). Design of Haptic Interfaces for Therapy. ACM Conference of Human Factors in Computing Systems http://portal.acm.org/citation.cfm?id=1518776. [12] Auger, J., & Loizeau, J. (2009). Smell+ http://www.auger-Loizeau.com/index.php?s=contact [13] Manabe, D. (2008). Face Visualizer http://www.daito.ws/work/smiles.html.

TO BLINK OR NOT TO BLINK

Alvaro Cassinelli and Stéphane Perrin

ABSTRACT A darkened corner reproduces the look of a small, gloomy apartment. A sofa placed right below the corner faces a TV set. The remote is visible on a small table that sits between the sofa and the set. This configuration naturally invites the spectator to sit and turn the TV on. Not noticing anything peculiar, they may browse some programs before briefly pausing on something. However, after less than ten seconds of watching, the TV will randomly switch itself to another channel. The spectator may switch back to the selected program using the remote, but there would now be a gap in whatever they were watching. If they do not use the remote, the new program will eventually be interrupted too, again in a seemingly random way. It will take two or three of these interruptions for the spectator to discover that they are causing these disruptions by blinking their eyes.

INTRODUCTION To blink or not to blink is a playful reflection on the active role that the (supposedly) passive spectator plays in postediting and censoring visual media. To give an example: it may take a whole day to generate five seconds of a stop motion movie; this means that blinking even for a fraction of a second equates to skipping several hours of hard work. Were the creators of the movie conscious of the possibility of this disrespectful intervention in the first place? Most likely not; but unless one straps the spectator to a chair, with their eyes artificially held open like Alex in A Clockwork Orange, blinking will be inescapable: it is a natural feature of seeing with human eyes. Blinking may not be the meanest form of disrespect that our body shows towards a filmmaker's hard work (going to the toilet in the middle of a movie is certainly worse), but it's symbolically more powerful: blinking randomly introduces a million tiny cuts in an otherwise carefully edited movie; it is a rain of razor blades that mutilate the film.

TECHNICAL STATEMENT The (technical) heart of the installation is the blink detection mechanism. This is achieved by continuously shooting a close up of the spectator's face. The camera is connected to a computer that processes the images in real time. After finding the face (using the OpenCV Haar-classifier library, integrated in openFrameworks), the location of the eyes is inferred by relying on standard face proportions (see Fig 1a).

46

Alvaro Cassinelli and Stéphane Perrin

TO BLINK **OR** NOT TO BLINK

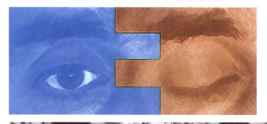

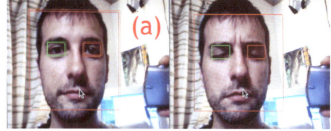

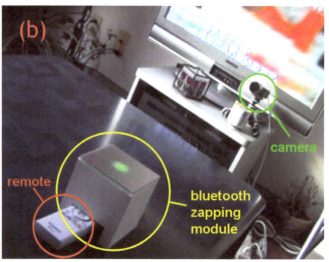

Alvaro Cassinelli and Stéphane Perrin

Figure 1. To Blink or Not to Blink installation setup.
[a] blink detection using openCV and Haar-classifiers (coded in C++ on openframeWorks).
[b] the setup including a camera, and the bluetooth zapping module that sits on top of an universal remote controller

Blink detection is performed by simple frame differencing in the region of the eyes. Efficient and robust detection of the blink is possible even in a dimly lit environment using an infrared light source. Each time a blink is detected, the information is sent wirelessly (Bluetooth) to a micro-controller (an Arduino Mini). The microcontroller then activates a small servo-motor that presses the buttons on a secondary, universal remote controller. This remote is inserted below the bluetooth zapping module (see Fig 1b) and both the remote and the zapping module are hidden somewhere nearby (for instance, under the table). The original remote controller is left visible on the table, perfectly accessible to the spectator. Even though the zapping module is invisible, the spectator can hear the characteristic robotic sound produced by the servo-motor as it presses the buttons on the remote. Since this sound perfectly correlates with blinking, it gives the spectator a direct clue about what is happening (incidentally, it produces an eerie feeling of being somehow part of the setup). A short demo video (1'56") can be seen at: http://toblinkornot.wordpress.com/.

DISCUSSION AND CONCLUSION To blink or not to blink seeks to amplify the consequences of blinking, so as to make them clear to the specta-tor. The idea is that the spectator will pay for these moments of fleeting disinterest by losing track of whatever they were watching. The spectator, made aware of this trade will be left with a dilemma: if they are truly interested in what they are watching, they will force themselves to keep their eyes wide open as long as they can (the only way to avoid disrup-tions producing significant gaps in the story). However, the more attention they put on the video, the less conscious they become of their natural body attitudes and they may blink inadvertently. If, on the contrary, they succeed in abstaining from blinking, they will still create gaps in the story, this time attentional gaps because of the effort that this unnatural attitude require. After a while, the spectator is forced to conclude that their organic body plays a fundamental role in modulating visual content and that as observer, they are neither innocent nor passive. They have to choose how they want to see the world. To blink or not to blink? After this experience, the spectator may look at the world with different eyelids existentialist eyelids so to speak.

ACKNOWLEDGEMENTS The authors would like to thank Jacqueline Steck for sharing initial code for blink detection.

Alvaro Cassinelli and Stéphane Perrin

EARLIDS & ENTACOUSTIC PERFORMANCE

ABSTRACT EARLIDS is a wearable device enabling the semi-voluntary control of auditory gain. Artificial earlids represent to the ears what natural eyelids are to the eyes: a fast and efficient reflex mechanism for protecting delicate sensory organs. Externally, the device presents itself as an ordinary pair of closed headphones, but hidden under each ear-cup we find EMG electrodes that monitor the contraction of the temporal and masseter muscles. When they contract or relax (consciously or unconsciously), the external sound being picked by binaural microphones placed on each side of the head will be greatly attenuated - or greatly amplified depending on the mode of operation. Applications may range from instant hearing protection without requiring the use of hands (in particular for people having to work at the boundary of environments with notably different sound levels such as night clubs), to the generation of highly personal acoustic experiences rendered possible by the manipulation of the environmental acoustic material.

INTRODUCTION The motivation for this experiment comes from the casual observation of small children performing a similar experiment, virtually always with a sense of awe. I remembered myself being scolded at least once without given a chance of explaining why I would enjoy placing my hands around my ears to hear the voice of the teacher change as if heard through a sea-shell, or rhythmically closing and opening the ear canals so as to brutally slice her (angry) words. The experience of a mute world that goes unnoticed to anyone but oneself is highly disturbing (the technique is routinely used in cinema as a way to render a scene dream-like, and as such relating more closely to one's own subjective world). There is a curious asymmetry with the sense of vision, perhaps precisely because we are endowed with eyelids and are thus habituated to the experience (for a related experiment, see 'To blink or not to blink' paper in this workshop). That such tampering of the senses alter the way the world is naturally perceived is obvious; autistic children and children with certain mental illnesses do something similar in order to self-stimulate and perhaps bring a certain regularity to their sensory input (the so called "stimming" behaviour that can include self talk, rubbing, jumping, but also tapping ears).

Alvaro Cassinelli

EARLIDS AS A SOUND MANIPULATION INSTRUMENT? The system proposed here explores the possibility of a highly individual auditory experience thanks to conscious or unconscious muscle tension and facial expression. Instead of using such information in an attempt to generate sound for others to hear as done by Manabe[1], we are interested in manipulating sound by the user of the interface, and for that user only. It is interesting to note that the body and its sensory organs always modulate the external sound field in one way or another. The ear transfer function is determined from a variety of parameters including the shape of the outer ear and internal hearing organs, head structure, middle ear muscle activity, etc. Neural low and high level processing also come into play to shape the final perceptual experience. However, if such subtle chain of processing is invisible and seemingly beyond the control of the person's mind, it is precisely because in normal conditions auditory sensor and motor integration produces the illusion of an objective, coherent, stable acoustic world that seems out there. The attempt behind this project is to amplify the effects of this unconscious processing to a level such that it will become readily apparent to the user that they are playing a crucial role in shaping the experience. We can then expect that, before the user re-learns the new artificial auditory sensory-motor contingencies, they will have the opportunity of consciously and playfully manipulate the raw-material constituted by the environmental sound. The importance given to this sensory-motor feedback makes this work depart from closely related experiments on audio augmentation such as those of Mueller[2] or Watanabe[3]. Taniguchi[4] describes a system capable of detecting discrete facial movements (e.g. sticking of the tongue) using infrared sensors embedded on the speaker's earbud. However, the system is used as an mp3 remote controller and it is not clear if it could be used to perform continuous modulation of audio parameters.

EARLIDS AS A PROTECTIVE DEVICE The human body presents a number of unconscious reflexes to protect the sensory organs from overly intense stimuli. The iris contracts as a way to limit the amount of energy that reaches the retina, but when this is not sufficient, eyelids provide a very efficient luminous barrier. Several muscles including muscles in the middle ear but also the masseter muscles contract naturally in human beings when exposed to loud sounds (the jaw acoustic reflex described for example by Deriu[5]). This natural reflex decreases the transmission

Alvaro Cassinelli

of vibrational energy to the cochlea, thus protecting the inner ear, but artificial earlids could dramatically increase the hearing dynamic range without the need of tapping ears with our hands.

EARLY PROTOTYPE In the early prototype presented here (see Fig 1-b,c,d), sound is first blocked almost completely by circumaural (ear-cup) headphones; sound is picked by bin-aural microphones and fed to a MAX/MSP patch on a laptop computer. The patch modulates the gain before redirecting the sound stream towards the headphones, using as a control signal the output of a custom made EMG detector based on an INA128 instrumentation amplifier. Integrating noise canceling technology, it would be possible to create a prototype in the shape of ear-buds or in-ear earlids (Fig 1-a).

DISCUSSION Sensual experiences can be brought about by the design of complex stimuli, or alternatively by tampering with the senses themselves. The first case is pervasive, and may well be the subject of a study in the Arts. The second case has, depending on the context, either positive or negative connotations. Tiredness, illness, and drugs can, as Huxley[6] put it, lower the efficiency of the cerebral valve so much that biologically speaking useless material flows into consciousness from 'out there', forcing the individual into close contact with naked reality (Appendix II).

Music is an example of a complex stimuli that requires specialized knowledge and training to produce. The skilled musician knows something about our senses, but indirectly: they know how one would react to each particular auditory stimulation. Complex, fined tuned machines called musical instruments are routinely designed to produce these sounds. On the other hand, devices that alter normal perception may bring extremely powerful sensual experiences not requiring the design of complex external stimuli; instead, by tampering with the senses, an otherwise crude flow of data (light, sound, etc) can be experienced as something enjoyable or of transcendental value. Molecular devices - drugs - have been used since time immemorial; more recently, electricity[7], magnetic fields[8], strobe lights[9] and more general audio-visual installations have been capable of producing similar effects in both controlled and uncontrolled environments for the purpose of scientific research, metaphysical inquiry or as means of pure recreational experience. Whether this sort of tampering of the senses is deceiving the soul, and is therefore of less intrinsic value than real art is a legitimate question, in particular if one standpoint is that art is something that can be somehow developed and perfected by an artist that is, to

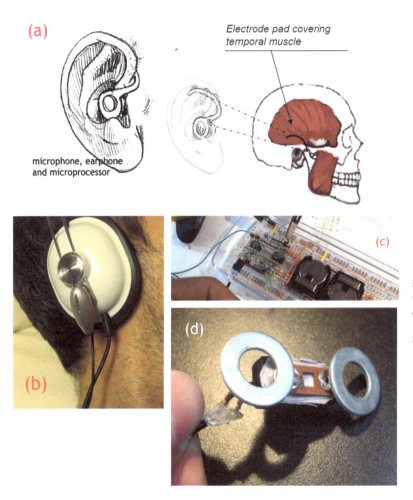

(a)

Electrode pad covering temporal muscle

microphone, earphone and microprocessor

(c)

(d)

(b)

Alvaro Cassinelli

Figure 1. Description of the EARLIDS system.

[a] wearable device with integrated EMG detector, microphone and earphone.

[b] Early prototype using closed headphones and external microphone.

[c] EMG detector based on INA128 instrumentation amplifier.

[d] temporal muscles electrodes (the reference electrode is worn on the wrist)

a certain extent, an actor and not a passive spectator of the whole experience. At any rate, the development of an art require skills, and offers the opportunity to explore perception in a systematic way. In this regard, and extending the concept of the EARLIDS, I propose to discuss a form of musical expression that requires skills and can be improved over an entire life, but whose concrete expression is intrinsically difficult to share - just like a highly personal drug-induced experience. While it is true that virtually all musical instruments bond to the body of the musician in more or less inextricable ways, the performance itself is an audible product for all to share, even though the sound itself may be relatively different for the performer an their audience (certain traditional instruments such as the jaw harp of Turkish tribes produce sounds that may be quite different from the point of view of the performer, because of strong sound bone transmission for instance), but both the externally audible performance and the subjective experience of the musician relate in a coherent way. Learning these differences suffices to cope with this disparity, and as a result the experience becomes interpersonal. In contrast, an entacoustic musician would be able to discuss at length a particular technique, but they could not directly share the resulting sensual experience.

CONCLUSION and further research This paper describes the principles of the EARLID, and introduces a prototype of the device, but most experimental research is yet to be done. If the device proves interesting, other entacoustic phenomena not necessarily originating in the ear may be amplified and integrated in the feedback loop. These may include heartbeat (to regulate tempo, for example), breathing, blood flow and reflow, etc. Again, these sounds may seem barely controllable (visceral functions are controlled by the autonomic nervous system without conscious intervention), but as the system purposely distorts the normal sensory-motor flow, it may force introspection and bring about some form of conscious control after some training (this of course is at the core of biofeedback techniques). Other applications of the system can be imagined, ranging from controlling the volume of an mp3 player or a mobile phone to more sophisticated functions such as control of the hearing gain as a function of the psychological state of the person (anxiety, relaxation, anger). A concrete example would be a system that detect stress, and reduces the auditory input in an attempt to soothe the person.

ACKNOWLEDGEMENTS The author would like to thank Daito Manabe, Carson Reynolds and Stephane Perrin for interesting discussions related to this work.

Alvaro Cassinelli

References [1] Manabe D. et al. (2008). Myoelectric sensors triggering sound, Retrieved Oct.6, 2010 from the World Wide Web: www. youtube.com/watch?v=r27KdzCgHT4. [2] Mueller, F. and Karau M. (2002, March). Transparent Hearing, CHI 2002 Extended Abstracts on Human Factors in Computing Systems, Minneapolis. [3] Watanabe, J. et al. (2006) Visual Resonator: Interface for Interactive Cocktail Party Phenome CHI2006 Extended Abstracts (pp. 1505-1510). [4] Taniguchi, K. (2009). Mimi Switch, Retrieved Oct.6, 2010 from the World Wide Web: www.engadget.com/tag/Kazuhiro+Taniguchi/ [5] Deriu, F. et al (2007, April) Origin of sound-evoked EMG responses in human masseter muscle. In: The Journal of Physiology, 580, (pp. 195-209). [6] Huxley, A. (1954, 1956). The Doors of Perception and Heaven and Hell, Harper & Brothers Ed. [7] Penfield, W. (1975). The Mystery of the Mind : A Critical Study of Consciousness and the Human Brain. Princeton University Press. [8] Khamsi, R (2004). Electrical brainstorms busted as source of ghosts, BioEd Online. [9] Geiger, J (2003). The Chapel of Extreme Experience: A Short History of Stroboscopic Light and the Dream Machine.

EXTRA ROOM: WHAT IF WE LIVED IN A SOCIETY WHERE OUR EVERY THOUGHT WAS PUBLIC?

Gunnar Green and Bernd Hopfengärtner

ABSTRACT Extra Room is a design project combining scale models and narrative elements to reflect on how advances in neuroscience and technology might affect our self perception. The Extra Room exists in an imaginary world where mind-reading technologies are used to read the human mind. As the mind becomes transparent in this world a new necessity of protective self discipline emerges. Utilising effects of sensory deprivation and methods used by the military to break someone down, the room enables subjects to adjust their thinking and beliefs. The Extra Room, a reversed disciplinary architecture, is built into the basement of a multi storey building where it is shared by the house's inhabitants.

INTRODUCTION When we started working on the project we noticed an increased public awareness in the subject of neuroscience. It was more appealing to look at the public reaction to the topic than dealing with the particular developments within the science itself. How would all the newly acquired knowledge about the workings of the human brain affect the way we perceive ourselves? Would we fall back into a functionalist understanding of the human, which already developed in the 19th century? In summer 2009 we noticed that the topic got more and more picked up in the press. Wired Magazine published an article about a conviction for murder in India based on brainwave patterns that apparently prove experiencial knowledge when showing photographs of the crime scene to the suspect[1]. At the same time a story about the usage of MRI scanners for job interviews spread across the blogs. A Dutch scientist claimed that using the technology in five years time would be possible and practicable[2]. Apparently the acceptance of using and believing the results of neuroscience research found more common ground. But how would these technological possibilities transform our culture, and the way we think about ourselves and about others? And centrally how can we approach and explore this topic as designers? We decided to look at the very extreme end: What would happen if mind reading was possible?

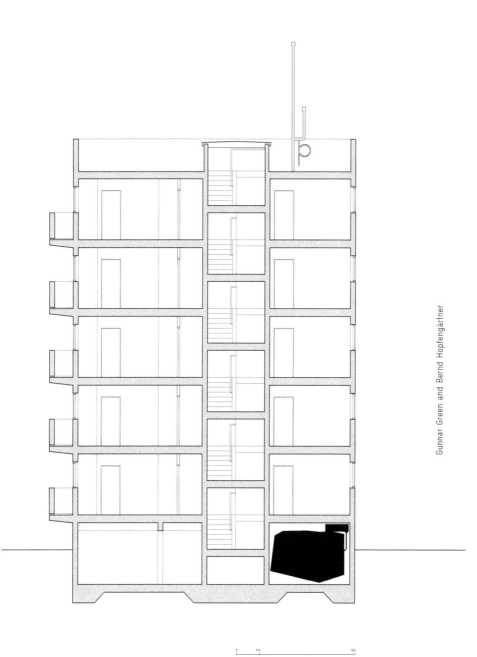

Gunnar Green and Bernd Hopfengärtner

Gunnar Green and Bernd Hopfengärtner

MIND READING Sociologist Erwing Goffman compares the socially acting individual to an actor on stage. Resuming his book The Presentation
of the Self in Everyday Life, he writes: '[...] the performed self was seen
as some kind of image, usually creditable, which the individual on stage
and a character effectively attempts to induce others to hold in regard to
him'[3]. In social interaction, we always present particular aspects of ourselves, we take on certain roles according to the context and other people
involved in the social situation. The technological possibility of mind reading would create a much more absolute and static conception of the self.
Independent of site or manners, if we were working or amongst our friends,
we would have to stay true to our self, cementing what we are and what
we will be. How could we adapt to new situations, how would we develop
our personality and how to combine career and private life? Extra Room
offers one solution, which is a process to alter ones thoughts and beliefs
according to the requirements of your actual situation.

CONDITION OF A MIND ALTERING ENVIRONMENT We designed a
room and a process people would subject themselves to, in order to
achieve desired psychological alterations. The Extra Room[fig.1] is an amal-
gamation of a sensory deprivation unit, an interrogation room and a prison
cell. The dimensions are 3m by 4m, it is soundproof, carpeted, has no
window and only one door[5]. The walls are tilted, having no horizontal
or vertical wall to prevent any spatial orientation and to induce mental
unrest[6]. The process itself is inspired by experiments of sensory deprivation. Sensory deprivation is the entire suppression of the senses and has,
as shown by Jack A. Vernon, a particular effect: After spending 24 hours
in a completely dark and soundproof room, people who were subjected to
propaganda showed to be eight times more susceptive compared to the
control group[7].

This result lead to the possibility in our project of using the effect for the
purpose of self induced mind control. The procedure would be as followed:
A person starts their session in complete darkness. During the whole
time liquid nutrition can be accessed through The Valve[fig.2] on the wall
inside the room. We choose liquid nutrition to eliminate the experience
of time, a consequence itemised food such as sandwiches would have.
The Extra Room also provides a toilet. After 36 hours with no stimuli the
room lightens and at this stage the person's mind would be hungry for
new input. A speaker, The Prompter, cites the desired alterations that had

been programmed in the room before entrance. This process of repetitive recitation lasts for 12 hours and ends with the opening of the door. We imagined this Extra Room in the basement of multi storey building where it is shared by the house inhabitants, both to give it something secret and to make it economically believable.

OF OTHER SPACES Extra Room is neither a prediction nor a proposal for a psychological strategy for the scenario of omnipresent mind reading technologies. Instead we aimed to create an analogy that refers to modern attempts to control the individual in psychological experiments, or in military and architecture. Looking for an adequate visual language, which expresses the structural violence that would emerge in a situation where technology becomes potent enough to dissolve the barriers between our internal processes and our impression to others, we got inspired by architecture. The Trellick Tower in London is a textbook example for brutalist architecture. With its 31 storeys, and separate elevator tower, it became an icon for this architectural movement and in fact for a functional conception of the modern human. In the Extra Room design, we used architecture as a medium describing the cultural and psychological conditions of our scenario. Distorted in shape, the Extra Room is hidden away in the basement of a multi storey building and is both a shared space for all the inhabitants and a space that integrates in but doesn't belong to daily life. As Michel Foucault writes: 'There are also, probably in every culture, in every civilization, real places [...] which are something like countersites, a kind of effectively enacted utopia in which the real sites [...] are simultaneously represented, contested, and inverted. Places of this kind are outside of all places, even though it may be possible to indicate their location in reality.'[8]

We can find many of these spaces and sometimes they become cultural icons like the garden shed in England, the fallout shelter in Switzerland or the sauna in Finland, all of them Extra Rooms telling specific stories about the cultures they exist in.

FUTURE WORK So far the Extra Room project consist of a video, scale models and photographs. In this form it has been exhibited as part of the What if ... exhibition at the Science Gallery Dublin[9] and at the Designed Disorder exhibition at the Center for Urban Built Environment in Manchester[10]. We see the context of galleries and exhibitions as the appropriate field for Extra Room to engage with an audience.

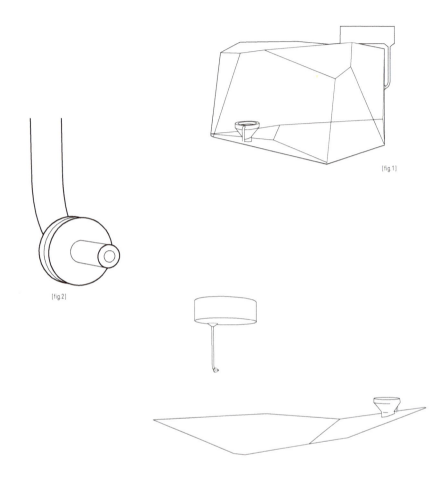

[fig.1]

[fig.2]

Gunnar Green and Bernd Hopfengärtner

References [1]Saini, A.: The brain police: judging murder with an MRI, WIRED, http://www.wired.co.uk/wired-magazine/archive/2009/05/features/guilty?page=all.[2]http://www.nextnature.net/2009/02/brain-scan-replaces-job-interview-in-5-years/.[3]Goffman, E.: The presentation of self in Everyday Life, Penguin books, 1990, p. 244.[5]CIA Human Resource Exploitation Training Manual, 1983 http://www.gwu.edu/~nsarchiv/NSAEBB/NSAEBB27/02-07.htm.[6]Murphy, M.: Glimpses of a future architecture in Miessen, M.: Did Someone Say Participate? Frankfurt: Revolver, 2006, p. 77.[7]Vernon, J. A.: Inside the Black Room, C.N. Potter, 1963, p. 39.[8]Foucault, M.: Of other Spaces. 1967 (Heterotopias).[9]http://sciencegallery.com/whatif.[10]http://www.andfestival.org.uk/event/designed-disorder.

TENSED UP - A PIECE OF MATERIAL DEMONSTRATES OUR FIELD OF ACTIVITY

ABSTRACT This paper describes a wearable sensor to feel, detect and indicate electricity. I want to combine material behaviour with human perception to enable communication and to raise specific questions regarding increasing fields of electronic technology and our electrified behaviour. A woven textile uses electrical energy from its surroundings via influence—by human activity as well as electric fields nearby—and passes it in a comprehensible way to the user. For testing it, the textile is attached at the shoulder of the participant and has exposed yarns, which is to represent hair. If he or she is acting fast, the textile hair stands higher and higher—it charges up until it wants to discharge in its surroundings. If the material received a huge quantity of electric energy, it gets more inflexible. After that it wants to give up its electricity and consequently can interrupt technical devices or give the wearer small electric shocks, after he charged it.

INTRODUCTION In medical technology electricity was used to divine interventions for diseases such as epilepsy. Today people know that all natural structures are based on electric processes, but what does it feel like? I used a muscle stimulation device to get a sense of electricity and to create expressions, to transform my body through electric current—like Daito Manabe or Stelarc did this before. It was a vague feeling—tickling, prickly and strained.

My material creates this tingling feeling by little electric discharges to the body. In general electricity evokes fear by this blurred sensation. Today there is skepticism about electromagnetic fields. 'All devices emit background signals—electrostatically, magnetically, acoustically, and optically-that are characteristic of particular devices. It could increase paranoia'.[1]

RELATED WORK In public vicinity, electricity cannot affect biology on the cell level. 'There are no broadly accepted long term methods to cause cancer'[2] depends on mobile phones or power lines. The fear of electromagnetic

Clemens Winkler

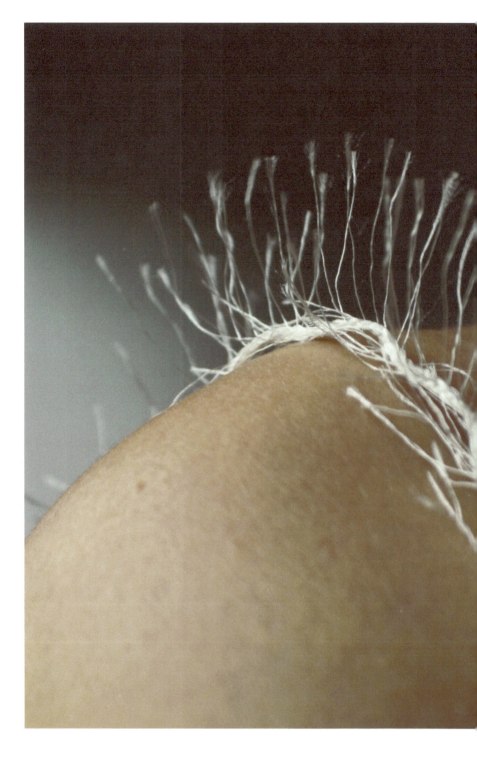

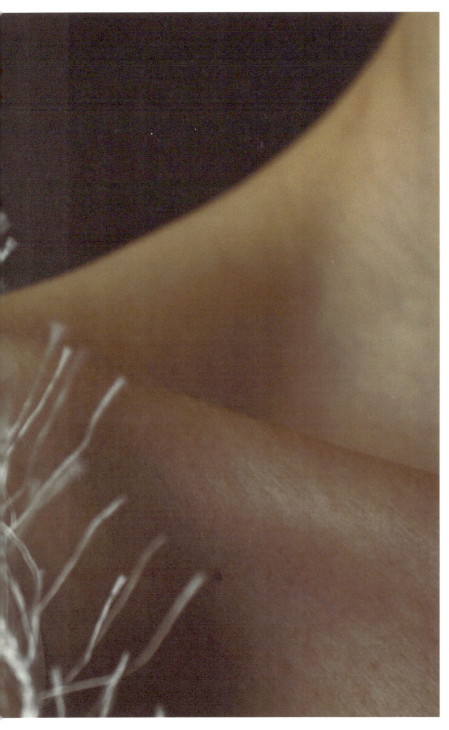

New Prototype (in progress)

fields has inspired artists in creating products (p.e. Dunne and Raby). Commercial EMF detectors respond to low frequencies in the range of 50 to 1000 Hz, but it is often unclear on which sources it depends.

DESIGN Firstly, I built headphones which I wanted to couple to other electric sources via an antenna and two coils. I made electric/electromagnetic fields audible to understand from nearby 50 kHz 1 Mhz. I was hearing clicks and snoozes, but still the fields remained abstract and seemed to be widespread. Then I designed the textile as an electrode antenna, that responds in an abstract way to our electrically charged surroundings. I designed a woven fabric, that functions as a power receptor as well as an object of visual and tactile expression for human awareness. Electricity gets perceivable by moving hair and flowing currents. Hair stood on end like in the hysteria in history and in todays topics about electric smog.

MATERIALIZING ELECTRONICS The embodied textile is woven by conductive and synthetic nonconductive yarns, capacitors and diodes. The received current will amplify via batch induction in a coil like in the charging unit of a camera flash it creates high voltages in conductive yarns and carries charges in textile. In the first steps I had to use a lot of electronic logic—a processor to define electric range and a large unit of capacitors, coils and diodes. Depending on patent of Nicola Tesla from the 19th century about collecting radiant energy, I transfromed electronic logic to material performance. So I only used a few capacitors and diodes within the textile. I wanted to find a material specific and imminent character. Here I used the possibility to changes the shape of material by electric charges as a force in every non-conductive material.

DISCUSSION AND FUTURE WORK I am exploring to sense invisible informations that surround us for adding context. Based on my hearing tests I can work on further material expressions. The textile could maybe used to indicate electric informations through air. It can sensorial envelope human environment (offices) rely on other disciplines like architecture. Visible material reactions are not as precise as audible signals in our perception, but on the other hand more interpretable—It will be explored in future work. Now I work on a new method to build a new electrosensitive cloth with better conductivity and for formal aspects. My material probe acts like an integrated version of a device.

CONCLUSION In my project I search for a possibility to enhance and sensitise materials to explore a changing in perception. Figuratively the textile caricatures the fear of electric fields. The material probe describes electric current as something natural, which has different manifestations. Inconvenient electric charges penetrates skin and technical devices in our closer surroundings. This fabric can ask questions about cultural trends that will emerge from our constantly growing need for energy. And the material How far can we empathize with it? On questions 'What does technology look like?'[3] I think technology must fuse with irrational in order to be useful. I want to choose the focus more on human-machine experiences, in particular our (dis-)like of certain common electric/material technologies.

Bottom: Prototype, Feb 2010

Clemens Winkler

References [1][2] Vaucelle, C. Ishii, H. and Paradiso J.A: Electromagnetic Field Detector Bracelet. MIT Media Lab (2008) [3] Coelho, M. Pulp-Based Computing: A Framework for Building Computers Out of Paper. In the International Conference on Ubiquitous Computing (Ubicomp). (2007) – Berzowska, J. and Coelho, M. Kukkia and Vilkas: Kinetic Electronic Garments, in the International Symposium on Wearable Computers (ISWC'05). IEEE (2005). – Dunne, A. and Raby, F.: Design Noir: The Secret Life of Electronic Objects. Birkhäuser Basel, 1 edition. (2001) – Fried, L: Social Defense Mechanisms: Tools for Reclaiming Our Personal Space, submitted to the Department of Electrical Engineering and Computer Science (2005).

OUTRE-RONDE,
COUNTER-INTUITIVE
SIGHT DILATION

Anne-Sarah Le Meur

Beyond-Round (Outre-Ronde) is an interactive 360 degree panorama that explores the action of the viewer's gaze, the sensitivity of the peripheral visual field and the resistance of an image to being seen. Stillness, slowness, micro-movements allow the viewer to progress towards contemplation. If the viewer moves as he/she usually does —faster than a certain slowness— they will experience diverse types of perceptive losses: the image will disappear under their gaze. If he/she does move more slowly than usual, the image will remain visible and acquire other behaviors. This counter-intuitive process (vision loss versus the power of slowness) makes the viewer more aware of their own behavior-desire of vision. Because their surroundings have become dynamic, reacting to their way of moving/perceiving, it gives them a sense of responsibility about how/what they are looking at/for and initiates in them the patience to construct a dialog, a relationship.

Beyond-Round has been influenced by a number of interactive art pieces[1] but more especially by the book of Tao[2] and the work of Samuel Beckett (Film, 1964, where Buster Keaton keeps on avoiding the camera gaze, and some of Beckett's writings dealing with darkness[3]). Not to deny my love of Rothko's paintings, James Turrell's light installations and Stan Brakhage's abstract films.

The visual part of the work, Eye-Ocean (Oeil-océan, 2007), constitutes the first stage of this research on vision: it consists of two virtual lights that move and reflect on a surface that is itself oscillating over a black background. One of the lights has negative parameters that render it black. They both move over and seemingly in the surface (in depth, as well as laterally), their movement making their sizes and relative distances change, one light sometimes swallowing the other. The whole phenomena remain abstract, minimal, albeit organic and almost sensual. I inserted in my code many parameters and aperiodic loops to generate and control the infinite visual variations in time. Some of those parameters trigger others, interfere with them, changing their sense of variation, as if the complex clock mechanism I developed would become crazy. To play with programming language and numbers is a core element of my creative process.

To experience Beyond-Round, the viewer stands and moves inside a cylindrical screen. A luminous image appears, disappears, circulates around him/her on the screen. Its speed and colors vary according to the viewer's behavior, its location according to the direction of his/her gaze. The image appears, in particular, in the peripheral visual field, an imprecise but very sensitive area for movements detection.

A simple light in darkness, the image becomes a very strong stimulus. If the viewer, perceiving the image in their peripheral vision, turns to center the vision, the image disappears, either in place or behind the viewer. If they slowly rotate, or stay still, phenomena become sustainable, even shift towards them (tamed?), and color range increases. Once in a face to face interaction, micro-movements modulate color intensity (among other things).

Importantly, observers can sit around the interacting viewer without interfering with the participant's peripheral visual field. Observers can see the entire screen from beneath. They thus see more than the participant, for example when the image appears behind the viewer.

There are 4 different phases to the work:

PHASE 1 Awakening: the still form gradually disappears when the viewer looks at it, or moves slowly toward them if they stay still. After 3 disappearances, phases 2 and 4 are possible.

PHASE 2 Fast corner: if the viewer turns fast in darkness, the image appears randomly at the edge of his/her visual field, moves faster in the same direction, vanishes. The speed of the image varies according to iterations (repetitions) and to the viewer's speed. Phase 3 is possible.

PHASE 3 Tango phase: how the image shifts and the length of its disappearance vary according to how long the viewer stands still before moving and to how many times they repeat their movement (iterations). It is possible to dance with the image in the periphery (as a slow tango) if the viewer moves very slowly. Phases 2 and 4 can follow this.

PHASE 4 Face to face interaction: The image is tamed, accepts to be seen by a direct gaze. Micro-movements trigger a range of attributes in the lights such as movement, glaring, color range, and darkening of form and its sliding, etc. Phases 2 and 3 are possible.

Anne-Sarah Le Meur

The viewer interacting with the mobile image, 'Fast corner phase'.
Beyond-Round seen from the inside.

The viewer interacting with the mobile image, 'Face to face interaction'.
Beyond-Round seen from the outside.

During a recent residency (Générale en Manufacture, Sèvres, France, September-October 2009) where I opened the work to the public, I was able to observe children and assessed that bifurcations between different phases were important to create the feeling of a very sensitive living being and dynamic environment. In each phase, the viewer's behavior can trigger variations (by repeating or changing their behavior) inside the current phase and change of phases. The desired moment, phase 4, Face to face interaction, is always possible to access except in the early stages of Phase 1, Awakening, and throughout the whole of phase 2, Fast corner. A distinction of Beyond-Round is that the more one moves (to see), the less one sees. The less one moves, the more one relates with phenomena. Slowness and speed allow different types of interaction and perception, all of them awakening the awareness of the limits of the visual field and of the power of the interaction. Diverse interacting stages are built up, with increasing subtlety (the form is increasingly sensitive - reacting), so that the viewer feels, understands and gradually adapts his/her behavior to the image behavior. Being an object of desire, the image teaches the viewer how to behave in order for an encounter to happen. He/She learns not to move, not to obey/fulfill his/her desire to watch if he/she wants to see something. In a way, the viewer experiences an initiation process.

Visual and kinesthesic perceptions are enhanced by a non-verbal colorful dialog in slowness, darkness, silence. They enrich the viewer's awareness of reciprocal and respectful relationship, to space and duration, and trigger psychic refocusing.

ACKNOWLEDGEMENTS Into the Hollow of Darkness, which includes Beyond-Round and Eye-Ocean, has been developed over ten years and comprises sensor prototypes, bespoke software and corporeal experiences. The project has received technical support from Francis Bras (Interface-Z[4], France), Didier Bouchon (Le Cube[5], France), and The ZKM[6] Institute for Visual Media in Karlsruhe, Germany.

NOTE ON THE VIDEO DOCUMENTATION To document Beyond-Round is not easy due to a number of factors including: darkness (camera sensitivity/additional lights needed to record), stillness/slowness (low dramaturgy), ghostlike appearance of the image, and circular space (the image is often outside the camera frame). Moreover, the intimate pleasure to see the image in the corner of the eye, the pleasure to wait, and rotating with the image are difficult (impossible?) to represent. It is necessary to live the interaction to assess its sensitivity and meaning.

Anne-Sarah Le Meur

References [1]Joachim Sauter and Dirk Lüsebrink, Zerseher, 1992, see http://vimeo.com/7560243 - Simon Biggs, Shadows, 1993, see http://www.littlepig.org.uk/installations/shadows/shadows.htm.[2]Lao-Tseu, Tao-Te King, Gallimard, 2002.[3] Samuel Beckett, Pour finir encore et autres foirades, Minuit, Paris, 1991; Samuel Beckett, Compagnie, Minuit, Paris, 1985. [4]http://www.interface-z. com/.[5]http://www.lesiteducube.com/.[6]http://www.zkm.de/
http://aslemeur.free.fr/projets/outre_r_eng.htm

Anne-Sarah Le Meur

ORGANIZER BIOGRAPHIES

CARSON REYNOLDS is a Project Assistant Professor at the University of Tokyo. His research interests are sensor systems, privacy, and roboethics. He holds a Doctor of Philosophy from the Massachusetts Institute of Technology from the Program in Media Arts and Sciences where he performed research at the Media Lab's Affective Computing group.

ALVARO CASSINELLI is Assistant Professor and leader of the Meta-Perception Group at the Ishikawa-Oku laboratory. His research interests lie in the area of human-computer interfaces for enhancing human communication and expression. He also enjoys working at the boundary between Science and Art. Alvaro Cassinelli was born in Uruguay and obtained an Engineering degree in Telecommunications and a Ph.D. in Physics while living in France.

TOMOKO HAYASHI is a multidisciplinary artist/designer trained in Japan and the UK. Her work explores the use of tactile and digital media to enhance the intimacy and human connection facilitated by contemporary communication technologies. She is currently working at The University of Tokyo in Meta Perception Group.

DANIELLE WILDE has an MA in Interaction Design from the Royal College of Art, London and is completing a PhD in body-worn technologies and the poetics of embodied interaction, at Monash University, Melbourne and CSIRO, Australia. She is currently a Visiting Researcher at Tokyo University in the Ishikawa Komura Laboratory Meta Perception Group.

ALEXIS ZERROUG is an assistant at the Ishikawa Komuro Laboratory at the University of Tokyo. He holds a Master's degree in Virtual Reality from ParisTech-Laval. Previously he received an engineering degree from the five-year program at ESIEA (Laval, France). His expanding interests include new human interfaces and new media art.

JUSSI ÄNGESLEVÄ balances between education, research and industry, holding guest professorship at the Berlin University of the Arts and working as Art Director at ART+COM design agency. Ranging from novel interaction design research to physical installations and architectural media, his work exists in publications, prototypes and patents as well as high profile installations in public space.

MATT KARAU is an Electrical Engineer (BSc. MIT, 2001) and educator whose professional experience runs the gamut from independent media art installations to electronics design for Fortune 500 companies. He has worked in the MIT Media Lab, the Media Lab Europe, and was a founding member of the Distance Lab. His teaching includes presenting courses in the No. 2 High School (EFZ) at ECNU (China), Trinity College Dublin, and the University of Design (HfG) Karlsruhe. Matt currently works for ART+COM and lectures in the University of the Arts (UdK) Berlin.

SUSANNA HERTRICH works as independent artist and designer in Berlin. Her body of work comprises photography, installations, machines to storytelling and video. She likes to challenge boundaries between Research and Art. Former affiliations to research facilities include Intel Corporation's People & Practices Group and Meta Perception Group of University of Tokyo.

INDEX

Index

www.ingramcontent.com/pod-product-compliance
Lightning Source LLC
Chambersburg PA
CBHW041143050326
40689CB00001B/469